Simplified Models for Assessing Heat and Mass Transfer in Evaporative Towers

Synthesis Lectures on Engineering

Simplified Models for Assessing Heat and Mass Transfer in Evaporative Towers
Alessandra De Angelis, Onorio Saro, Giulio Lorenzini, Stefano D'Elia, and Marco Medici
2013

The Engineering Design Challenge: A Unique Opportunity
Charles W. Dolan
2013

The Making of Green Engineers: Sustainable Development and the Hybrid Imagination
Andrew Jamison
2013

Crafting Your Research Future: A Guide to Successful Master's and Ph.D. Degrees in Science
& Engineering
Charles X. Ling and Qiang Yang
2012

Fundamentals of Engineering Economics and Decision Analysis
David L. Whitman and Ronald E. Terry
2012

A Little Book on Teaching: A Beginner's Guide for Educators of Engineering and Applied
Science
Steven F. Barrett
2012

Engineering Thermodynamics and 21st Century Energy Problems: A Textbook Companion
for Student Engagement
Donna Riley
2011

MATLAB for Engineering and the Life Sciences
Joseph V. Tranquillo
2011

Tensor Properties of Solids, Part One: Equilibrium Tensor Properties of Solids
Richard F. Tinder
2007

Essentials of Applied Mathematics for Scientists and Engineers
Robert G. Watts
2007

Project Management for Engineering Design
Charles Lessard and Joseph Lessard
2007

Relativistic Flight Mechanics and Space Travel
Richard F. Tinder
2006

Simplified Models for Assessing Heat and Mass Transfer in Evaporative Towers

Alessandra De Angelis, Onorio Saro, Giulio Lorenzini, Stefano D'Elia, and Marco Medici

ISBN: 978-3-031-79359-2   paperback
ISBN: 978-3-031-79360-8   ebook

DOI 10.1007/978-3-031-79360-8

A Publication in the Springer series
SYNTHESIS LECTURES ON ENGINEERING

Lecture #22
Series ISSN
Synthesis Lectures on Engineering
Print 1939-5221   Electronic 1939-523X

Simplified Models for Assessing Heat and Mass Transfer in Evaporative Towers

Alessandra De Angelis
University of Udine, Italy

Onorio Saro
University of Udine, Italy

Giulio Lorenzini
University of Parma, Italy

Stefano D'Elia
University of Parma, Italy

Marco Medici
University of Parma, Italy

SYNTHESIS LECTURES ON ENGINEERING #22

ABSTRACT

The aim of this book is to supply valid and reasonable parameters in order to guide the choice of the right model of industrial evaporative tower according to operating conditions which vary depending on the particular industrial context: power plants, chemical plants, food processing plants and other industrial facilities are characterized by specific assets and requirements that have to be satisfied.

Evaporative cooling is increasingly employed each time a significant water flow at a temperature which does not greatly differ from ambient temperature is needed for removing a remarkable heat load; its aim is to refrigerate a water flow through the partial evaporation of the same.

Often industrial processes require cooling machines or applications capable to remove the heat absorbed during working cycles. Evaporative cooling is the only transformation which is not directly implemented in conditioning systems and, facing high amounts of heat loads one needs to consider the presence of thermal sources which, in nature, act as best receptors for high energy fluxes: atmospheric air, rivers, lakes and sea water. Furthermore it is widely known that, given equivalent thermodynamic conditions, water-cooled exchangers prove more compact and less costly than air-cooled ones.

Also, it is important to consider that the necessary quantity of natural water may not be always available for several reasons: physical absence of considerable amounts of water and presence of laws which safeguard the hydrologic environment are the most recurring circumstances that one has to face. In such cases the only solution is a system able to cool continuously re-circulating water. The evaporative tower is precisely the particularly efficient type of exchanger able to realize such a thermodynamic cycle.

KEYWORDS

evaporative towers, cooling machines, heat and mass transfer, zero-dimensional model, one-dimensional model

Contents

List of Figures

List of Tables

Preface

Engineers are not so often entrusted with the task of sizing evaporating towers, they are rather asked to select one according to market offer. In order to solve such a problem; builders supply clear reference diagrams to help customers in choosing the right model to match the desired operating conditions.

The aim of this work is not only to provide general documentation on the different types of evaporative towers for industrial use, and on their building and application criteria, but also to propose two mathematical models for assessing its global performance. The zero-dimensional approach outlines the operating conditions supposing a linear variation of saturation line, while the one-dimensional model presents a numerical analysis for determining the conditions for each cross section along the tower.

The reference book is divided into chapters developing the following topics:

- chapter 1: Description of commercial and industrial water cooling systems. Comparison of the various types considering the thermal issue, plant engineering, health and financial aspects.

- chapter 2: Typical applications of industrial evaporative towers. Description of installation and alteration criteria in pre-existing plants.

- chapter 3: Description of evaporative towers operating conditions in winter and in summer. Setting adjustment according to the external temperature or to the tower load.

- chapter 4: Description of the building criteria adopted for evaporative towers. Comparison between natural draught and forced draught towers. Focus on air intake through fans for the second type. Guidelines for selecting building material. The chapter is concluded with a building example by MITA group.

- chapter 5: Description of the operative principle of evaporative towers. In addition, a summary of the properties of humid air, the phase diagram capable to represent all the transformations that humid air can undergo and notes about some exchange principles are provided.

- chapter 6: Focus on water treatment issues. Incrustation, mud and corrosion caused by lack of analysis of water physical-chemical conditions. Available solutions, including preventive and chemical cleaning systems, are shown.

- chapter 7: Description of the zero-dimensional model for determining the conditions of air and water when leaving the tower. Comparison of the results obtained with the Merkel model, the zero-dimensional model and field measurements.

- chapter 8: Description of zero-dimensional model application using electronic sheets. Three calculation cases for water and air conditions when leaving the tower. Comparison of the results with the bulletins drawn up by a supplier.

- chapter 9: A one-dimensional approach is proposed for studying the conditions of water and air inside an evaporative tower. Temperature, enthalpy and humidity variation for each cross section is determined through numerical analysis.

- chapter 10: Description of Euler and Runge-Kutta methods for solving the one-dimensional model. Comparison of pros and cons of these two solution types.

- chapter 11: Description of one-dimensional model application through VBA. Analysis of results and comparison of calculated and measured values.

Alessandra De Angelis, Onorio Saro, Giulio Lorenzini, Stefano D'Elia, and Marco Medici
April 2013

CHAPTER 1

Evaporative Cooling

1.1 INTRODUCTION

Economic development has enabled industries to greatly develop. The drawback is they now have to cope with the issues connected with the disposal of production process heat waste. Such problem affects companies which produce steam, refine petrol, deal with mechanical parts cooling or with exothermic reaction control, or use a refrigerating system or air conditioning technologies.

In case of commercial and industrial air conditioning system, facing small amounts of heat-to-remove, the required capacity of refrigeration amounts to a maximum of few kW and it is sufficient to consider HVAC (heating, ventilating, and air conditioning) evaporative towers. Otherwise, in case of heavy industrial processes with large amounts of heat-to-remove, the most suitable choice is to adopt one or more industrial cooling towers, capable to refrigerate tens of thousands of cubic meter of water per hour. Air can be still used for removing limited amounts of heat, as the production volume of the majority of plants is high and therefore they generate huge quantities of heat, the only efficient way allowed by law for disposing of them implies the use of massive amounts of water. For those reasons, generally industrial evaporative towers are much larger than HVAC towers.

Over the last years, closed-loop systems characterized by water circulating through a cooling tower have hence become the most common solution even if evaporative towers were originally reserved to big users, as more moderate loads were easily satisfied by privately owned water wells or by waterworks. However, the present difficulty in providing small users with natural sources has called for increasingly undersized, lighter and affordable towers. The market currently offers cooling towers that even meet size and economic requirements of extremely small users.

1.2 COMMERCIAL AND INDUSTRIAL REFRIGERATION: HVAC

An HVAC cooling tower is used to dispose of heat by using a chiller. In terms of energy use, water-cooled chillers are normally more efficient than air-cooled chillers: air-cooled chillers must reject heat at the higher dry-bulb temperature, and thus have a lower average reverse-effectiveness. Liquid chillers are machines featuring a refrigerating loop for cooling water or other liquids.

Using a chiller, the temperature of the cold water output does not depend upon ambient conditions; hence, it is obvious that the output of chillers is better, more accurate and safer. Large office buildings, hospitals, and schools typically use one or more cooling towers which adopt this kind of closed-loop.

Basically, the HVAC use of a cooling tower is equivalent to a cooling tower with a water-cooled condenser, even if with different features due to the different size. To avoid wastage and the disposal issue, cooling water is continuously circulated around an hydraulic loop by means of a circulating pump.

1.3   INDUSTRY: WATER COOLING METHODS

Both production and economic optimization have definitely become a true priority in nowadays industrial sector. Such priority has generated the need for cold water for simplifying, improving and speeding up the industrial production process. On the other hand, industrial plants must now cope with increasingly strict laws and directives, whose aim is to contain the wastage of a resource which has become scarcer and to control the way polluting water waste is disposed of (as even excessively hot water damages the environment). Thus, also for the industrial cases closed-loops are adopted and in this circumstance they are rather strictly necessary. In such systems a single amount of water undergoes a continuous heating and cooling cycle, eliminating pointless wastage and discharge issues that especially in the industrial field would be of particular relevance. The closed-loop water cooling systems available for industrial use adopt liquid coolers that cool water using ambient air; in this case the temperature of the cold water output is strongly bound to ambient conditions. In comparison with the liquid chillers used in HVAC systems, it is true that the output water is characterized by worse quality but on the other hand one must consider that chillers are complex devices with extremely high initial installation costs and they are also characterized by rather high energy consumption and maintenance costs. For these reasons, when operating conditions allow it, simpler and more affordable systems, like the liquid coolers, are preferred: for instance, when the cold water produced must reach a temperature which is higher than ambient temperature almost all year long (spent oil cooling process in plastic mould presses) or, when the cold water requirement is so huge, like in the case of the steel industry, to make chillers absolutely uneconomical.

1.3.1   EVAPORATIVE TOWERS AND DRY COOLERS

Machines using ambient air to cool liquids are evaporative towers and dry coolers. The main feature which distinguishes such categories is the means employed for transferring heat between such two fluids. In traditional evaporative towers, heat transfer occurs through direct contact between air, propelled by a fan, rising along the tower and water, which is sprayed inside the upper part of the tower, falling in the opposite direction of air and then collected at the bottom. Such method is based on the difference between air wet-bulb temperature and water temperature. Heat transfer is very efficient, as part of the water evaporates once it comes in contact with cold air: evaporation subtracts a remarkable quantity of heat from the water (latent heat of evaporation). Water temperature equal to, and sometimes lower than, ambient air is thus reached.

In dry coolers the heat transfer between air and water takes place by employing finned coils, which are conceptually identical to those used for the condensation stage in air condensed refrigerating cycles.

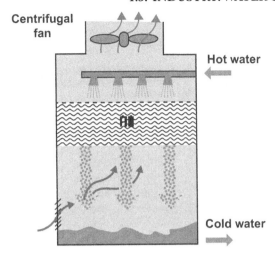

Figure 1.1: Evaporative tower: operating diagram.

Fans propel the air through the finned coils cooling the water circulating inside the piping. In this case, water and air do not get in touch; the disadvantage is poorer transfer efficiency (difference between dry-bulb air temperature and water temperature) matched with all the advantages peculiar of a fully closed hydraulic loop. Water minimum temperature is usually 2–3°C higher than that of ambient air.

The efficiency of ambient air liquid coolers strongly depends upon the temperature rise deriving from heat transfer between air and water. Compared with dry coolers, evaporative towers offer a slight advantage considering the degree of heat transfer efficiency. Another disadvantage connected to towers is the constant water loss (evaporation can be assessed between 5 and 10% of the treated volume) which must be promptly counterbalanced by additional water. Double-loop cooling towers are another option. In such systems there are two separated hydraulic loops (the primary, which carries water towards the supply point and the secondary, for cooling) which are connected by means of a heat exchanger.

1.3.2   HEALTH CONSIDERATIONS

Yet, direct air-water contact, which is the reason of such advantage, is the cause of the many disadvantages found in evaporative towers if compared to dry coolers. Direct contact between air and water promotes algae growth, triggers incrustation and oxidization of the metal parts of the hydraulic loop; this situation must be kept under control with anti-algae, anti-corrosion additives and continuous maintenance.

Pathogen proliferation (Legionellosis, for instance) inside the tower is another problem, as it represents a risk of infection for those who inhale the vapors released by the process. A team

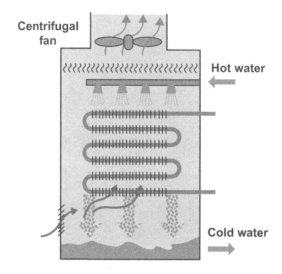

Figure 1.2: Dry cooler: operating diagram.

of researchers found that Legionella bacteria travelled up to 6 km through the air from a large contaminated cooling tower at a petrochemical plant in Pas-de-Calais, France and the consequent outbreak killed 21 of the 86 people who had a legionella infection. In order to avoid health risks, in some countries, among them Australia, the use of evaporative towers is strongly restricted by law. On the contrary, dry coolers are not affected by algae growth, incrustation, oxidation and pathogen proliferation. As air and water never get in touch, the water does not contain the oxygen necessary for the development of such phenomena. Therefore, the degree of performance of dry coolers is certainly slightly inferior to that of cooling towers but the former imply definitely lower running costs and are much simpler to operate.

1.3.3   INSTALLATION COSTS

Considering installation costs, the evaporative towers generally make the most affordable option. Yet, such initial financial benefit, which at any rate is remarkable only in the case of large plants, is then rapidly nullified by much higher running costs. In a nutshell, cooling towers become an advantageous option only when the water flow to be treated reaches dimensions able to increase advantages such as lower installation costs and a higher degree or efficiency.

CHAPTER 2

Evaporative Towers Applications

2.1 TYPICAL APPLICATIONS

Following, some of the most interesting applications of industrial evaporative towers (hereinafter called evaporative towers) taken from the several involving water cooling for industrial purposes are considered:

- steelworks, foundries, petroleum industry;

- vinegar factories, forges, chemical baths;

- engine stands, brickworks, malt-houses;

- paint factories, conditioning systems, diesel engines;

- air and gas compressors, oil-mills;

- steam and gas condensers, production plants, atomic piles;

- frozen food industry, dry cleaners;

- alcohol and fat distilleries, chemical industry, printing works;

- rubber works, pharmaceutical industry, glassworks.

In many of the cases included in the above list, evaporative towers are employed for cooling and recycling water which is directly used as a cooling agent (for instance, in steelworks for cooling parts of the electric furnace, the ingot moulds and the mill; in internal combustion and diesel engines; for cylinders in air and gas compressors; for water and oil emulsions used for removing shavings in mechanical processes, etc.).

In many other cases evaporative towers are used for cooling water which is used as a cooling agent in heat exchangers or in condensers (for instance, malt cooling in breweries, cooling of induction furnaces in foundries, condensation of water vapor in fat industries or in canneries in general).

2.2   PRODUCTION PLANTS

Among the above mentioned applications, we must point out the role that evaporative towers play in cooling, and hence recirculating, the water used for removing heat from condensers in air conditioning and industrial refrigeration units (food storage systems: cold rooms for meat, fruit and vegetables, ice-creams, and frozen food; freeze driers and systems for cooling liquids: systems for wine, beer, soft drinks; ice production; artificial ice rinks and all the other multifarious uses of artificial ice).

According to cooling agent type, refrigerating and conditioning systems can be divided into three categories:

1. Systems which exclusively employ groundwater as direct cooling agent (for air conditioning only).

2. Systems featuring two cooling agents: groundwater and evaporating refrigerant fluid. Groundwater, sometimes used as cooling agent, can also be employed for cooling the condenser of the evaporating refrigerant fluid.

3. Systems that achieve the cooling effect by totally depending upon evaporation of a refrigerant fluid which is recondensed inside an exchanger cooled by air, groundwater or by thermally regenerated water from a cooling tower.

2.3   PLANNING NEW PLANTS

The best condenser cooling system for a new plant is chosen bearing in mind the following criteria:

• financial: considering purchase and installation costs and respective depreciation, not forgetting masonry works for adapting the system to the premises;

• economic: considering running costs including: electricity consumption by compressors, pumps and fans, tower make-up water cost, maintenance, painting, disincrustation operations, possible make-up water treatment, cost of groundwater (extraction costs plus taxes);

• law: according to state and regional laws regulating the use of groundwater and of state-owned water, or to laws regulating ambient noise level in the type of area (industrial or residential) where the plant stands.

In the great majority of cases, the best option for new installations is a condenser cooled with water treated by a cooling tower (as a matter of fact, a tower costs less than digging a 25/30 m deep well, or also than special materials for the condenser in the case sea water were employed).

2.4   ALTERATION OF PRE-EXISTING PLANTS

Systems employing groundwater often have to undergo alteration once such source is no longer available due to depletion, to measures enforced by the official authorities, or to excessive cost. The following methods are adopted for replacing well water as cooling agent:

- installation of air-cooled condensers;

- installation of a cooling tower or of an evaporative condenser, which is a special type of cooling tower.

If on one hand water consumption, and related costs, are significantly lower in such systems than for the groundwater option, it does not mean that the following consequences can be avoided:

- rise in energy consumption to power the compressors;

- decrease of the degree of refrigerating performance of the system;

- increase of the ratio between condensation and evaporation pressure with consequent fall of the operating conditions of the compressors, especially if of the reciprocating type.

The fall of the degree of refrigerating performance is not an important factor and, in most cases, does not exceed the planning tolerance forecast. Whereas, the rise of energy absorption is much higher and costly and may also imply the need to replace the electric engine control unit of the compressors or to reduce the speed of the latter.

Air-cooled condensers are a good solution for low power consumption plants, provided that: climatic conditions are suitable; water is virtually absent; operating periods are short or are exclusively performed in winter. This type of condenser proves much less affordable, if compared to evaporative towers or condensers, in the case of large plants given the increased size, weight, and hence cost of the devices, as well as the rise in absorbed energy and in fan noise rate.

Furthermore, in the case of pre-existing plant conversion, the refrigerating loop must be thoroughly altered to eliminate the old water condensers.

Finally, condensation temperatures obtained with air cooling systems are much higher than those achieved by evaporative towers or condensers and this makes the drawbacks illustrated above even worse when compared to groundwater systems, as condensation temperature will be 5–6°C higher than ambient temperature, which exceeds wet-bulb temperature by 8–9°C.

The only strong point featured by air condensation systems is that they do not employ water: a quality which may be essential under particular conditions.

Evaporative condensers are devices which also contain the pipes in which the refrigerant fluid condenses and operates according to the same principle governing evaporative towers. The cooling water is pumped from the water collection sump into the nozzles, which spray it upon the pipes in which the refrigerant fluid condenses. The condensation heat of the refrigerant fluid, transferred to the water by the pipe wall, makes part of the water evaporate; hence heat is transferred to the air. Fans provide the degree of air circulation necessary for the process.

Evaporative towers are more popular than evaporative condensers in pre-existing systems for the following reasons:

- with evaporative condensers, refrigerant fluid piping must be altered and pre-existing water condensers cannot be used;

- evaporative condensers are more expensive than towers;

- evaporative condensers are subject to more severe corrosion attack.

Yet it must be borne in mind that systems adopting evaporative condensers are simpler. Cooling towers allow reducing water consumption rate by 2%–10% in comparison to an open system, and also achieving conditions which are more suitable for specific technological processes, avoiding the problems that low-temperature groundwater may create (for example, cylinder cooling in air compressors or stationary diesel engines).

CHAPTER 3

Evaporative Towers Installation

3.1   GENERAL CRITERIA

The installation of an evaporative tower does not generally imply particular problems. The best setting for each tower may be selected according to the following:

- the tower must be positioned to allow correct air intake thus avoiding air recirculation inside the same;

- the tower must not be positioned in the proximity of heat sources or of river discharges, especially containing even an extremely limited quantity of corrosive or polluting products;

- droplets carried outside by the outgoing airflow may generate humidity throughout the area surrounding the tower; in addition, with strong wind and within a certain distance, such humidity could turn into snow or ice whenever severe freezing conditions occur;

- tower position, installation and type must be accurately considered in areas subject to strict noise level regulations;

- even though evaporative towers are not usually particularly heavy, they can be installed on roofs or terraces only following due assessment of the bearing structure resistance and any necessary reinforcement work. Load transfer of a tower ranges from 500–1000 kg/m^2, according to type of building;

- a stand is required for ground installation (a platform or supporting beams) to ensure the tower position is levelled. As a matter of fact, fan rotation axes must be placed correctly for perfect operating conditions and to avoid unexpected strain;

- as regards the hydraulic loop, the plan must fulfill all the criteria which are usually followed for minimizing load leaks; of course the nearer the tower to the devices which employ the treated water, the simpler, and hence cheaper, the loop. Such condition is achieved by minimizing the use of bends and by selecting pipes of suitable diameters.

3.2   WINTER OPERATION

Many applications require towers to operate throughout winter. The main issues connected to such season are: water in the collection sump may freeze during periods of inactivity and ice may form on the air intake grids and between the fan impeller and the ring within which it rotates, while water spray may generate ice in the area surrounding the tower.

To avoid water freezing in the sump, the same can be fitted with an electric or steam heater. Another option is using an auxiliary sump, to be installed at a lower level of the tower where freezing does not occur, as an alternative. Such auxiliary sump must be sized for:

- collecting all the water discharge from the tower or the pipes when the system is stopped;

- guaranteeing adequate pump suction power;

- filling and emptying the loop piping without causing pump priming loss.

The most efficient measure for avoiding ice formation around the tower in winter is certainly switching off the fan. The result is that droplets are no longer drifted, strong condensation does not occur and part of the outgoing water vapor, which is very humid and hotter than ambient air and becomes saturated and oversaturated as soon as it cools leaving the tower, does not freeze.

With low external temperature, the heat load can generally be disposed of also without resorting to the fan, by natural convection and by conduction.

The fan can be stopped manually or by adopting one of the systems mentioned in the "Temperature adjustment and control" chapter.

If the heat load the tower must handle is so high to make it impossible to reach the desired temperature for outgoing water temperature without operating the fan, the latter can be activated with a double pole engine: by setting the lowest speed, ice formation and electricity consumption will be reduced.

3.3   TEMPERATURE ADJUSTMENT AND CONTROL

Cooling towers must be sized to suit a potential able to dispose of the maximum heat load possible in summer months, with peak wet-bulb temperatures. If no adjustment is performed, the tower may produce water cooler than average during cool seasons or at night.

In some particular cooling processes, water temperature must not drop under certain values; the fan or the water circulating inside the tower can be adjusted to suit them. The simpler system is to operate the fan intermittently, switching it on and off manually. Whereas the best option is often to adjust the temperature of cooled water automatically, according to external temperature or to tower load, as manual operation may not be sufficiently prompt or accurate. Automatic adjustment is commonly achieved by:

- stopping the fan by means of a thermostat installed inside the collection sump each time the temperature of outgoing water is too low. If such method is chosen, fan stop and start temperature (thermostat differential) must be selected with utmost care to protect both the fan and the engine from mechanical strain and stress deriving from excessive switch-on and switch-off;

Figure 3.1: Natural draught cooling tower (photographer: Petr Kratochvil, http://www.publicdomainpictures.net/view-image.php?image=342&picture=torri-di-raffreddamento).

- stopping the fan by means of a thermostat installed in contact with ambient air, often in parallel with the previous, each time freezing conditions occur;

- installing a mixing valve and a short circuit pipe between incoming hot water and cooled water leaving the tower. If the temperature of the cooled water falls, the thermostat controlled outgoing water mixing valve short circuits part of the water entering the tower, mixing it downstream with the water leaving the same. As soon as the temperature reaches a satisfactory level, the thermostat mounted on the supply point outlet pipe stabilizes the mixing valve. Yet, the amount of circuited water must not exceed a third of total flow, allowing the cooling tower to operate with a flow equivalent to at least two thirds of normal flow, for constant irrigation of the fill material.

- installing a solenoid valve and a short circuit pipe between hot incoming water and cooled water leaving the tower. Such system, which is very similar to the one described in the previous point, allows less accurate, yet still efficient, adjustment. The solenoid valve, mounted on the short circuit piping, is controlled by a thermostat whose bulb is installed on the outlet pipe carrying water to the supply point. If the temperature of the cooled water decreases, the solenoid valve short circuits part of the water entering the tower, mixing it downstream with the cooled water leaving the same. For short circuited water flow restriction, see the previous point;

- adding and mixing fresh water. Cooling towers are generally sized according to the ambient air conditions which occur during summer months (medium-maximum wet-bulb temperatures). When such conditions are exceeded for the duration of some hours or some days, and it is not possible to achieve the normal temperature that cooled water should reach in that particular period by just employing the tower, the correct temperature can be obtained by adding cold bore or aqueduct water. In most cases it is sufficient to adjust the fresh water flow manually. The ideal option for saving water is an automated system employing a motorized modulating valve installed on the cold water adduction piping and controlled by a thermostat immersed in the water collection sump. It must be stressed that the temperature that the water mixture can reach depends on the ratio between the water circulating inside the tower and the quantity of added cold water, as well as on their temperature. For instance, by mixing 1000 litres of 32°C water with 100 litres of 15°C water, the outcome will be 1100 litres of water whose temperature will be approximately 30°C.

3.4   CAPACITY ADJUSTMENT AND CONTROL

If a cooling loop serves several supply points, it is advisable that the quantity of water circulating through the tower is always kept constant, even when one or more heat exchangers are excluded. If the exclusion of one or more exchangers stops the corresponding amount-part of water flow, the hydraulic load of the tower will continuously vary, leading to imperfect irroration of the fill pack.

It is therefore important, when possible, to keep the quantity of tower inlet water under control, short circuiting water circulation in the excluded exchanger. Such condition can be achieved by automatic adjustment, applying the following methods:

- by installing, on the cooled water loop and just before each exchanger or supply point, a three-way modulating valve controlled by a thermostat whose bulb is positioned on the inlet of the primary loop of the exchanger. When all supply points are excluded, and if circulation is not completely interrupted, the whole water flow is directed through the short circuits. The result is constant water flow inside the cooling tower;

- by installing adjustment valves and an intermediate sump. In such capacity adjustment systems the cooling loop includes a "tower loop" and a "supply point loop." The water collection sump is divided into a basin for cold water and one reserved to hot. They are connected by an overflow valve so that, when water level in one of the two basins exceeds surge level, water flows back into the other one. The hot water returned by each supply point is pumped back into the tower. The water cooled by the tower falls into the cold water sump, from which it is pumped toward the various supply points. The amount of water required for each supply point can hence be adjusted, manually or automatically, without altering the tower loop settings. If one or more supply points are excluded, loop pressure will increase and excess water must be directed into the cold water basin by means of a pressostatic valve. A thermostat can be used for controlling the engine of the hot water pump or that of the tower fan (as suggested in the "Temperature adjustment" paragraph);

- by feeding the supply points by means of an autoclave and adding an intermediate sump. Such solution is recommendable when many intermittent operation devices with extremely variable concurrency factors are connected to the same loop. This option includes a sump divided into two sections, one for hot water and one for cold. A pump takes the hot water (returned by the supply points) from the first section and pumps it into the tower, for the cooling process. Such water then flows into the cold water section. Water is then pumped and directed from the latter to an autoclave featuring almost constant pressure above water level. A thermostat placed on the cold water sump controls tower fan operation. When all supply points are inactive, which is when no water is returned to the hot water sump, the hot water pump is stopped, too. Cold water pump operation is controlled by a water level-sensitive indicator placed on the autoclave. The process must not be interrupted and started more than six times per hour.

CHAPTER 4

Evaporative Towers Building Criteria

A series of multifarious devices are needed for building evaporative towers which operate according to the evaporative tower principle. The following criteria are used for classifying each tower:

1. evaporative sumps

    - basic

    - with spray nozzles

2. natural draught towers

    - open or atmospheric

    - with or without fill material

    - hyperbolic

3. mechanical draught towers

    - permanent, with wall bearing structure, for great potentials

    - packaged pre-assembled.

4.1 SUMPS

Evaporating sumps make the simplest technological application of the evaporative cooling principle: inside them, water circulates slowly and is cooled by the evaporation process which occurs on the open surface of the sump. Such system obviously only achieves a low cooling effect and very large sumps are needed for obtaining sufficient heat transfer surface.

Procedure efficiency can be enhanced by installing nozzles for spraying water above the collection sumps: in this way the contact surface between water and air is greatly increased.

4.2 NATURAL DRAUGHT TOWERS

The next stage in the development of evaporative cooling was achieved with open or atmospheric towers: towers with wall shutters, which allow air to circulate freely. Water is sprayed by a series of nozzles placed in the upper part of the tower and then falls, in the form of droplets, into the collection sump. During the fall such droplets come in contact with the air circulating within the tower. Yet, in this method, the potential of the tower strongly depends upon wind intensity and consequent droplet drift: water loss increases as wind does. Part of the air also gets sucked into the tower by induction caused by the falling water. The degree of performance of such type of tower can be improved, and droplet drift reduced, by employing small wood beams or other material for filling the gaps. Further progress was made by building the first closed-wall towers based on air circulation triggered by the natural draught which occurs inside the same.

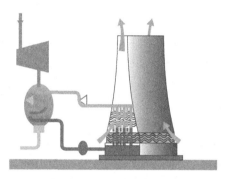

Figure 4.1: Natural draught tower.

The air present in the tower is heated and humidified by the hot water, its specific gravity therefore decreases and air consequently tends to rise and leave the tower, attracting other air which undergoes the same physical transformation and generating a permanent current. The difference between air density inside and outside the tower produces a sufficient level of draught if the tower is adequately tall.

Lower and more basic towers are made of wood, while larger ones have concrete walls.

Finally, natural draught hyperbolic towers are used in plants with an hourly intake of thousands of cubic meters and low temperature rise. Such towers must be obviously considerably tall for providing good chimney and cooling effect. Base diameter in hyperbolic towers measures approximately 50 m and height reaches 80–100 m.

4.3   MECHANICAL DRAUGHT TOWERS

Given their operating limits, the previously examined systems are only used in particular situations. The degree of performance of natural draught towers remarkably falls in summer and with warm or mild climate, as the extremely slight difference between tower internal temperature and that of external air reduces the chimney effect, and hence the quantity of air circulating inside the tower. Mechanical towers are therefore much more popular: among them, towers with concrete walls are mainly reserved to higher potential plants.

Pre-assembled evaporative towers, or types which can be easily assembled on the site, are the best option for the majority of applications. Such towers may contain fill material able to increase water-air contact surface. In the case fill material is not present (nowadays a quite rare case, required only for particular processes), the droplet surface acts as the actual water contact area. Air velocity must be kept low (2–2.5 m/s) and it is necessary to install an efficient droplet separator to avoid heavy water drift.

Towers containing fill are generally the most popular option.

Water distribution can be achieved by employing a series of overflows, by forcing water to fall through perforated sumps or by using a set of spray nozzles: all means for separating water into droplets thus increasing the water-air contact surface. The nozzle solution is the most used, even though it increases water mechanical drift in the air current. Another means for distributing water is the two-arm-shaft Segneri device. The water that flows out through small holes along the arm's generator causes a reaction which makes the system rotate and the water fall on the fill pack.

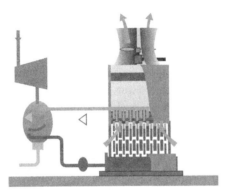

Figure 4.2: Mechanical draught tower.

Air movement can be classified as crossflow or counterflow, while water usually falls vertically. Counterflow towers feature smaller bulk surfaces, while crossflow towers may be lower.

The majority of builders chose the counterflow solution.

It is a particular type of induced draught tower that achieves air drifting through the ejecting effect produced by water sprayed by high pressure nozzles in the contracted section of an approximated Venturi tube.

4.4   FANS POSITION AND TYPE

A fan assures correct air circulation in mechanical draught towers; such device may be placed downstream or upstream of the fill material: the latter is therefore positioned in air current either blown or sucked by the fan.

Downstream positioning was the first option to be adopted, as it was a simple way of attracting air circulation in natural draught towers during the hottest periods.

Upstream positioning is a more recent invention and was adopted for avoiding that the heaviest devices, namely engines and fans, had to be borne by the facility, which would have had to be reinforced. With such method the fans (bulk, shell or stand) could become part of a much lighter structure.

Towers with upstream fans become even more important once the need for plants of greater capacity increased; in that case radial fans were employed to lessen noise level.

The two different solutions are realized with both axial and radial-flow blowers. Yet, radial fans placed downstream of the fill material, hence in suction current, are only used in smaller plants (up to 10 cubic meters of water). Axial fans are suitable for more remarkable water flows.

Several manufacturers prefer the solution with axial fans placed upstream of the transfer area, yet such arrangement is not the most widely adopted as such fans, if positioned downstream with equal load, would absorbed less energy. It must be stressed that cooling towers require fans to provide heavy flows with little static pressure. Radial fans perform the worst under such conditions (performance close to that of open outlet). Whereas large double suction low rotation radial fans (300–400 rpm) are more suitable. The problem is that such ultra silent fans are expensive and heavy; therefore, smaller fans performing 500–700 rpm are normally preferred for reducing the weight load.

Yet such devices require higher power supply and are noisier, reaching the level of the best axial fans. On the other hand, axial fans are the best choice for heavy flows and low pressures, hence they perform well and noise level is acceptable. Axial fans operate within such type of application with performances classified as 0.45 or even 0.5 while the radial ones do not exceed a performance equivalent to 0.2 and often even much lower.

Such a low degree of performance adds to the total loss of dynamic pressure, which in the case of upstream fan arrangement is purposely nullified at tower entrance for achieving uniform air distribution in the fill pack, so it is no surprise that the consequent power requirement of a tower fitted with radial fans is two or three times that of an identical tower with axial fans. In addition, axial fans can nowadays achieve noise levels which are competitive if compared to those obtained with small radial fans; if, in addition, axial fans operate not exceeding 300–450 rpm, the noise produced is so low they can be used in residential areas.

4.5   CORROSION ISSUE AND MATERIAL SELECTION

The other great issue involving cooling towers, leaving aside air circulation fans, is corrosion. Corrosion attacks each of the elements which make up a cooling tower with a different level of severity; prevention measures must therefore be varied.

The bearing structure frame of assembled towers should be preferably made of heat or hot-dip galvanized steel framework, which must be sufficiently thick and, in the case of aggressive environmental conditions, varnished with a suitable protection. The thickness and shape of the framework must guarantee excellent zinc coating also in the holes. Stainless aluminium alloys, previously anodized or protected with chrome based varnishes, may be employed in particular cases. Panelling does not involve great technical, as well as aesthetic, problems and stainless steel and reinforced polyester are the only materials which provide satisfactory solutions. Galvanized sheet, leaving aside any aesthetic consideration, is too unpredictable for being considered sufficient, especially on the long term. E-coat (normally epoxy) bake varnishing provides a good solution which has been widely experimented for car bodywork.

As regards materials for building axial fans, which are immersed in the humid air current, the best option, and largely used on ships, are aluminium-magnesium alloys (which are also light, a quality which is extremely important in very large plants).

A varnishing cycle is sufficient for preserving water collection sumps realized with metal sheet at least 3 mm thick if they are accurately maintained. Stainless steel is too expensive and hot-dip galvanizing would deform them, while polyester is an excellent option for modular sumps featuring smaller elements (8–10 m^2 each).

Some types of plastic are ideal for all the other parts which make up a tower, namely water distribution pipes, nozzles, droplet separators and fill material. This is because such parts undergo extreme conditions involving continuous wetting and drying cycles. A potential weak point of plastic material is that it may catch fire if it comes in contact with a naked flame, a situation which is unlikely in cooling towers. As matter of fact, they are set outdoors and in hardly accessible positions and such danger is usually very remote. The event has actually occurred only in cases in which the oxy-fuel flame used for welding pipes was carelessly placed near a plastic surface. Hence, the problem is avoidable and is to be exclusively ascribed to imprudence and inexperience.

The market offers several towers bearing a maximum flow of 350–400 m^3/h, featuring shells entirely made of reinforced plastic, a solution which obviously protects also the body of the tower from corrosion. However, if budget is not an issue, it is sufficient to replace the heat exchange surface, which is generally the largest part made of flammable material, with a non-flammable one without resorting to metal, which is more vulnerable to corrosion as well as heavier than plastic.

Accurate material planning for each part is useful for dramatically reducing, even if not eliminating, danger deriving from corrosion and to keep weight load and bulk under control.

It is obvious that also the whole material selection process must consider cleaning, water treatment, scaling encrustation and anti-algae treatment, as well as the cost of spare parts [1].

4.6 SAMPLE MODE

The following model of a cooling tower is taken from MITA group production (Figure 4.3).

A brief description of the components:

1. MOTOR-FAN

The axial motor-fan unit includes:

- three-phase electric engine (IP 55 protection, tropicalized winding, multi-voltage and multi-frequency voltage);

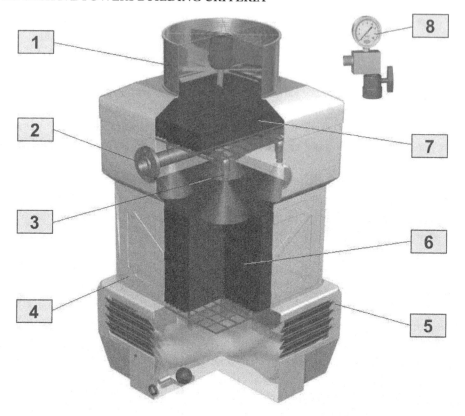

Figure 4.3: Description of components: 1. axial motor-fan; 2. water distribution pipe; 3. spray nozzles; 4. tower body; 5. sump; 6. fill pack; 7. droplet separator pack; 8. hydrometer unit – pet cock valve; (www. mita-tech.it).

- steel engine support ring, hot-dip galvanized after processing;
- axial fan directly coupled to the electric engine, with aluminium hub (models: PMS 65-85-110-130-180-240-280-360) or steel hub (model: PMS 260) and interchangeable blades;
- AISI 304 stainless steel fan protection grill manufactured pursuant to current safety regulations.

2. WATER DISTRIBUTION PIPING

Made of plastic material (PVC, POLYPROPYLENE or POLYETHYLENE according to specific application), it is constituted by a main manifold and side branches holding the spray nozzles. A flanged inlet connection is present.

3. SPRAY NOZZLES

Static type nozzles are made of polypropylene according to MITA design. They feature large water outlets to avoid blockage and guarantee perfect fill pack irroration with full cone spray.

4. TOWER BODY

The free-standing shell of the tower is fully made of glass-reinforced plastic and is hence corrosion and maintenance free.

The external surface is totally protected by first class isophthalic gelcoat, purposely conceived for guaranteeing protection against U.V. rays. The lower part of the shell, named "body," houses the fill pack, the water distribution piping, the glass-reinforced flap guards and, usually, includes the sump with all the hydraulic fittings. The upper part of the shell, called "cap," is designed to support the motor-fan and contains the droplet separator.

5. SUMP

The glass-reinforced plastic sump is of a piece with the body. It contains all hydraulic fittings (cold water connection, make-up floater, overflow valve, bottom drainage) while the upper part is reserved to the glass-reinforced flap guards. The inside part is sealed with waterproof water repellent isophthalic paraffin gelcoat, for duration and inalterability over time without further treatments.

6. FILL PACK

The fill pack is the core of the machine. It must be constantly kept clean and intact to assure the full cooling efficiency of the tower. The standard version is made of PVC, using sheets made of vacuum thermoformed sheets which are then glued together and feature 19 mm wide gaps for air and water flow.

Height is variable according to the thermal rise it is meant to perform; the 100 mm upper layer is thicker for best endurance to dynamic strain caused by pressurized water sprayed by the nozzles. Special versions are realized with different configurations and materials, according to nature and/or temperature of water.

7. DROPLET SEPARATOR PACK

The droplet separator pack holds back the droplets avoiding their possible deviation outside by the air flow sucked by the fan. As the fill pack, it must always be kept clean and fully efficient to prevent pollution (spores, fungi or bacteria) in the humid air expelled by the tower.

The standard version is made of PVC, using sheets made of vacuum thermoformed sheets which are then glued together.

Special versions are realized with different configurations and materials, according to nature and/or temperature of water.

8. HYDROMETER UNIT—PETCOCK VALVE

This important accessory is composed by a meter water column calibration hydrometer in glycerine bath, with AISI 304 casing, a tap and a three-way connector.

In the absence of a flow measurer, such device allows, according to flow loss by the spray nozzles (equivalent to the pressure stated by the hydrometer hand during operation), to benefit from an immediate and well approximated circulating water flow check.

The plastic pet cock valve is used for simplifying water hardness testing or for draining the hot water piping manually.

CHAPTER 5

Operating Principle

The operating principle is known to everyone: the partial evaporation of a mass of water causes the remaining part to cool. The quantity of evaporated water normally ranges from 3%–4%, which means 96–97% of recirculated water is reclaimed. Common means for enhancing evaporation include forced air circulation with fans and interposition of materials with large surfaces and limited total volume (fill packs) which achieve maximum water-air contact. The practical applicative limit of cooling towers is the maximum water cooling temperature they can achieve inexpensively, which is 3–4°C above wet-bulb ambient air temperature. For example, with summer maximum ambient temperature of 32°C and 50% relative humidity (equivalent to 23.5°C wet-bulb) it is possible to achieve cheap water cooling reaching approximately 27–28°C. However, as such maximum temperature is reached only during short periods throughout a year, temperatures which are actually much lower can be achieved over a great part of its operation time.

Before describing the operating principle in detail, here is some general information on the thermodynamic parameters which come into play.

5.1 THERMODYNAMICS TECHNICAL NOTES

Evaporative towers are based on a particular way of exchanging heat, called mass transfer, which enables heat to be transferred to the air by means of a process which will be analyzed later.

Before focusing on how an evaporative tower works, it is time to summarize which properties humid air bears, to illustrate the state diagram on which all the transformations humid air may undergo can be represented and to revise a couple of heat transfer general principles.

5.1.1 FIRST LAW OF THERMODYNAMICS

Consider a closed system that exchanges with the outside, the elementary work dL and the elementary heat dQ during the time interval dt. The relation that expresses the first law is defined by introducing two other energy terms which characterize the system in its instantaneous constitution and conditions: the first, defined in relation to a space-time reference, is broken down into variation of kinetic energy dE_k and variation of potential energy dE_p. The second, which may be defined independently from any external reference, is called variation of internal energy dU. The relation reads $dQ=dE_k+dE_p+dU+dL$. If the closed system does not move respect to the inertial reference, it follows that it is $dE_k=0$, and, furthermore, if dE_p is fully ascribable to the gravitational field, it is also $dE_p=0$; hence the previous becomes $dQ=dU+dL$. U is a function of the variables necessary for defining the system, namely a system state function. Summarizing, the internal energy variation of a system is equal to the sum of the performed work (positive if carried out by the system, negative if underwent) and of the heat exchanged by the system (positive if absorbed, negative if released).

5.1.2   HUMIDITY

Air can be defined in different ways according to its composition:

- Atmospheric air, mainly containing nitrogen and oxygen, and a mixture of other gases, including carbon dioxide, water vapor and various polluting substances. It is the air we actually breathe, the one used in conditioning systems.

- Dry air, made up of nitrogen (around 78%), oxygen (around 21%) and of small quantities of gas, like argon, neon and carbon dioxide.

- Humid air, which is a mixture of dry air and water vapor.

To define humid air values one must consider that, while the gases (N_2, O_2, Ar, etc.) which constitute air are found far beyond their own critical temperature, hence they remain in the gas state, water vapor is instead found below its critical temperature and can become liquid (dew, fog) or even solid (frost). Dalton law is applicable to humid air: the total pressure is the sum of the partial pressures, that is, the pressure that each component would exert if it occupied the entire volume of the mixture:

$$p = p_{N_2} + p_{O_2} + p_{A_r} + \cdots p_{H_2O}. \tag{5.1}$$

The pressures of the components are proportional to their quantities expressed as the number of moles; in the case of water vapor we have:

$$\frac{n_v}{n} = \frac{p_v}{p}. \tag{5.2}$$

If more water vapor is added to a given volume of air, the partial pressure of the former increases. When such partial pressure reaches the value of vapor pressure at that temperature, the air is said to be saturated. At that point the water vapor starts to condense as a liquid with temperatures over the melting point, or as ice crystals (snow or frost) with temperatures under the melting point. The ratio between the partial pressure of the water vapor and vapor pressure at that temperature is called relative humidity. It must be stressed that air becomes saturated only if it comes in contact with liquid air (above or near a body of water, when it rains), otherwise it contains a quantity of water (in vapor state) which is lower than saturation. If it contains a higher quantity, it condenses and separates (fog, dew).

The water glass experiment clarifies the concepts of dry and of humid air (Figure 5.1). It is carried out by placing a glass full of water under a bell jar containing dry air at T constant temperature and p=1 bar pressure. After a while, part of the water contained by the glass evaporates and the vapor formed by the process transforms the dry air into humid air. The process continues until the mass mixing ratio reaches saturation point.

There are two possible values for defining the quantity of water vapor present in the air:

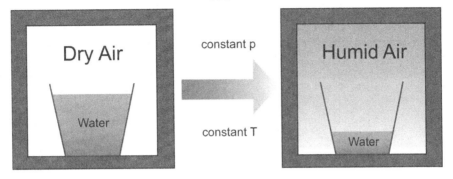

Figure 5.1: Water glass experiment.

- The *specific humidity* or humidity ratio x defined as the ratio between vapor mass and dry air mass in the mixture:

$$x = \frac{m_{vap}}{m_{air}}$$ (5.3)

- The relative humidity, defined as the ratio of the actual mass of vapor to the mass of vapor required to produce a saturated mixture at the same temperature:

$$\varphi = \frac{m_v}{m_{v,sat}} = \frac{p_v}{p_s}.$$ (5.4)

This value is included between 0 and 1 and can be defined as a percentage by multiplying it by 100.

The value of saturated vapor pressure depends upon the temperature; some of its values are reported in Table 5.1.

The values of saturated vapor pressure according to temperature can be represented graphically (Figure 5.2). In addition, it is possible to obtain an equation which approximates the experimental data of the table. In literature there are many correlations between saturated vapor pressure and temperature. A very accurate one, in the range from 0–200°C, is that of Hyland and Wexler (1983) [1]:

$$\ln(p_s) = \frac{c_1}{T} + c_2 + c_3 T + c_4 T^2 + c_5 T^3 + c_6 \ln(T)$$ (5.5)

where T[K] is the temperature at which p_s[Pa] is evaluated and, for $T > 273.15$ K,

$c_1 = -5800.2206$ [K]
$c_2 = 1.3914993$ [–]
$c_3 = -0.048640239$ [K^{-1}]
$c_4 = 4.1764768 \times 10^{-5}$ [K^{-2}]
$c_5 = -1.4452093 \times 10^{-8}$ [K^{-3}]
$c_6 = 6.5459673$ [–].

Table 5.1: Pressure and mass mixing ratio according to temperature

Air temperature t (°C)	Saturated vapour pressure p_s (bar)	Mass mixing ratio x (g_{vap}/kg_{air})
-20	0.00102	0.63
-15	0.00163	1.01
-10	0.00256	1.60
-5	0.00396	2.49
0	0.00600	3.78
2	0.00705	4.37
4	0.00812	5.03
6	0.00934	5.79
8	0.01072	6.65
10	0.01277	7.63
12	0.01401	8.75
14	0.01596	9.97
16	0.01816	11.4
18	0.02062	12.9
20	0.02336	14.7
22	0.02642	16.6
24	0.02982	18.8
26	0.03360	21.4
28	0.03778	24.0
30	0.04241	27.2
32	0.04753	30.6
34	0.05318	34.4
36	0.05940	38.8
38	0.06609	43.5
40	0.07358	48.8

A less accurate one but adequate for engineering purposes over 0°C is that used in the Standard ISO 13788, derived from the Clausius-Clapeyron equation:

$$p_s = 610.5 \cdot \exp\left(\frac{17.269 \cdot t}{237.7 + t}\right) \qquad [\text{Pa}] \qquad (5.6)$$

where t is the temperature of the humid air mixture, in celsius degrees.

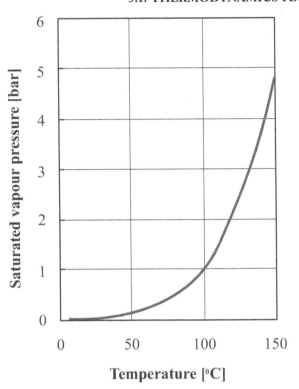

Figure 5.2: Pressure of saturated vapor according to temperature.

The relation between humidity ratio and relative humidity can be obtained from their definitions and from Dalton's law $\frac{n_v}{n_a} = \frac{p_v}{p_a}$, the result is that:

$$x = \frac{m_v}{m_a} = \frac{M_v n_v}{M_a n_a} = \frac{18.02 \cdot p_v}{28.07 \cdot p_a} = 0,622 \frac{p_v}{p - p_v} = 0.622 \frac{\varphi \cdot p_s(T)}{p - \varphi \cdot p_s(T)} \qquad (5.7)$$

in which n_v e n_a, respectively, represent the number of moles of vapor and of dry air present in the considered humid air mixture, and M_v and M_a are the molar masses of the two components. p is the total pressure of the considered humid air mixture and p_s is the vapor saturation pressure at the temperature of the mixture.

The value of x such that $\phi = 1$ is named saturation humidity ratio x_s. The temperature, at which the moist air becomes saturated at the same pressure p, with the same humidity ratio x as that of the given moist air, is called dew point temperature. When the earth's surface cools under such a point during the night, dew forms if it is above $0°C$, while frost appears if it is under $0°C$.

5.1.3 ENTHALPY

Enthalpy is a state function of a system expressing the quantity of energy that it is able to exchange with the environment. For example, in a chemical reaction the enthalpy exchanged by the system consists in the heat absorbed or released during the reaction. In the case of change of phase, for instance a liquid becoming gas, evaporation heat is the variation of the enthalpy of the system. While in a temperature variation process, the enthalpy is given by the thermal capacity under constant pressure. The definition of enthalpy H is:

$$H = U + pV \tag{5.8}$$

where U is the internal energy, p is the pressure and V is the volume of the system.

A reaction with negative enthalpy variation is exothermic, whereas a reaction with positive enthalpy variation is endothermic. In the case of a mixture the enthalpy of the system is the sum of contributions from each component; for the moist air as mixture of dry-air and water vapor:

$$H = H_a + H_v. \tag{5.9}$$

5.1.4 SPECIFIC ENTHALPY AND SPECIFIC HEAT OF AN AIR-VAPOR MIXTURE

Starting from Equation (5.9), the specific enthalpy of an air-vapor mixture is defined with reference to the mass (or the mass flow rate) of dry air:

$$h = \frac{H}{m_a} = \frac{H_a + H_v}{m_a} = \frac{m_a h_a + m_v h_v}{m_a} = h_a + \frac{m_v}{m_a} \cdot h_v \tag{5.10}$$

$$h = h_a + x \cdot h_v \tag{5.11}$$

in which h_a is the specific enthalpy of dry air, while h_v is the specific enthalpy of water vapor. Assuming for dry air $h_a=0$ at $T_0=273.15$ K ($t_0 = 0$ °C) as conventional state at zero enthalpy, and behavior of ideal gas with constant specific heat at constant pressure, the result is:

$$h_a = c_{p,a}(T - T_0) = c_{p,a}(t - t_0) = c_{p,a} \cdot t. \tag{5.12}$$

Assuming for water the conventional state at zero enthalpy of saturated liquid at the triple point $T_T = 273.16$ K. The specific enthalpy of water vapor, depending exclusively upon the value of the temperature, as considered ideal gas with constant $c_{p,v}$, can be calculated with this relation:

$$h_v = r_0 + c_{p,v}(T - T_T) \approx r_0 + c_{p,v}(T - T_0) = r_0 + c_{p,v} \cdot t \tag{5.13}$$

in which r_0 is the value of water vaporization heat at $T = 273.16$ K.

For the moist air mixture the specific heat at constant pressure c_p is the weighted average of the specific heat of dry air and of vapor for the given mixture. From Equation (5.11), its definition is:

$$c_p = \frac{dh}{dt} = c_{p,a} + x \cdot c_{p,v}. \tag{5.14}$$

5.1.5 PSYCHROMETRIC DIAGRAM

A thermodynamic state of moist air is uniquely defined if the total pressure and two independent properties are known. Usually, systems operate on moist air at barometric pressure that may be considered constant. A psychrometric diagram is a graphic representation of the proprieties of air in various situations. It is a representation of the thermodynamic state of moist air, at a constant total pressure p, traditionally, standard sea-level pressure has been used. Assmann realized the first psychrometric diagram, in connection with his invention, the psychrometer or hygrometer. The chart has a Cartesian coordinate system with dry bulb temperature t on the axis of abscissas, and humidity ratio x on the ordinate axis.

The diagram employs three different temperature measuring standards:

- dry-bulb temperature, which is air temperature measured with an ordinary thermometer. The scale is placed at the bottom of the graph and the vertical lines indicate equivalent dry-bulb temperatures;

- wet-bulb temperature, which is determined by forcing air above a thermometer wrapped in a wet cloth. It reflects the refrigerating effect of water evaporation, which causes a temperature lower than dry-bulb temperature. The scale is positioned along the curve in the upper left part of the graph and the inclined lines indicate equal wet-bulb temperatures;

- dew point temperature, namely the temperature below which humidity present in the air condenses. The scale is along the curve in the upper left part of the graphic and the horizontal lines indicate equal dew point temperatures.

The percentage of relative humidity is represented in the psychrometric diagram by means of curved lines, which originate from the lower left part and reach the upper right part. The line for 100% corresponds to the wet-bulb and dew point temperature scales. The 0% line is instead on the dry-bulb temperature scale.

Other psychrometric diagrams use enthalpy and humidity ratio as basic coordinates with many advantages. Constant wet-bulb temperature lines, dry-bulb temperature lines and a majority of psychrometric processes have a straight-line representation on $h–x$ coordinates. By mean of the relations presented in this chapter we may construct an $h–x$ chart. Experience has shown that a better result may be achieved when enthalpy is used as an oblique coordinate and humidity ratio as a vertical coordinate. This way the zone of unsaturated air becomes wider, then the graphical representation is made on a rectangular area with the intersections of dry bulb temperatures indicated on the horizontal line corresponding to $x = 0$. Figure 5.3 shows a psychrometric chart on $h–x$ coordinate system, with enthalpy as oblique coordinate to obtain the constant dry bulb temperature line of $50°C$ vertical on the right of the chart, as in the ASHRAE diagram (SI units).

Phase transitions that can be represented within the psychrometric diagram involve a latent heat flow between fluid and ambient air. The latent heat flow is the heat resulting in a change of moisture content occurring in a process without a change in temperature. The common form of

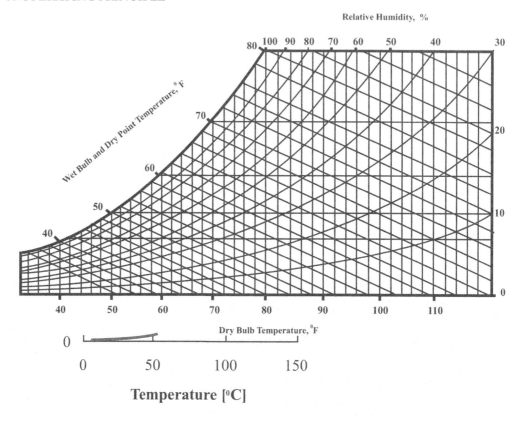

Figure 5.3: Psychrometric diagram.

latent heat described by the diagram is the latent heat of vaporization: the change in phase occurs from the liquid to gas, and vice versa. This process is associated to the phenomena that happen within an evaporative tower, precisely when all the fluids involved have the same temperature.

5.2   OPERATION OF EVAPORATIVE TOWERS

In evaporative towers heat transfer between water and air takes place not only by conduction and convection, but also and mainly through a process named mass transfer. The quantity of heat transferred from water to air is much higher with such method than with conduction and convection. Heat exchange by mass transfer, the classic process which is the basis of all evaporative coolers, may be simply explained as follows: air circulation promotes the evaporation of a small quantity of liquid, which absorbs the heat necessary to such phase change from the remaining water, which hence cools down. The phenomenon is extremely important considering that, to change from liquid to vapor state, a kilogram of water requires approximately 2500 kilojoule, which are equivalent to those needed to cool off 6°C as many as 100 kg of water.

Such a principle was also known in ancient times, when water was kept cool by storing it in porous baked clay amphorae, while varnished or decorated, hence waterproof items, were used for holding oil or wine. As a matter of fact, water can be kept cool in a porous container, as a small quantity of it filters through the same and, once it comes in contact with the atmosphere evaporates, keeping both the holder and the liquid inside it fresh.

Therefore, in cooling towers evaporation, and consequently heat transfer to air, is made possible by a difference between the pressure of the water vapor present in the thin layer of air directly in contact with the water (a thin layer which has the same temperature as the water and is saturated) and the pressure of the same vapor in the remaining surrounding air. Heat transfer can continue until the vapor concentration in ambient air equals the concentration of vapor in the thin borderline layer.

Such condition occurs when water temperature (and therefore that of the air of the borderline thin layer), equals the wet-bulb temperature of atmospheric air. At such point the pressure of the vapor in the air of the borderline thin layer, which is saturated, equals that of the vapor present in ambient air (bearing equal heat content) which is saturated, too; hence the pressure difference which determines heat transfer is lost. The wet-bulb temperature of ambient air is therefore the minimum temperature which may be in theory reached by water cooled through the evaporative process.

In this way water can be cooled under ambient air dry-bulb temperature (which is impossible with all the other transfer processes based on a temperature gradient), as wet-bulb temperature is always lower than dry-bulb temperature, except in the case when ambient air is saturated and the two temperatures are identical.

The potential of a tower depends on the thermal rise that water must undergo inside the same and on the difference between the temperature of the water leaving the tower and air wet-bulb temperature. In a given tower, the quantity of heat which can be transferred increases, with the same thermal rise, as air wet-bulb temperature decreases and that of the water entering the tower increases.

So, it is clear that it is essential to specify ambient air wet-bulb temperature to define the operating conditions of an evaporating tower. It has also been proven that the most severe conditions for evaporative towers occur in summer months, which are characterized by higher wet-bulb temperatures.

If a very high wet-bulb temperature is indicated when purchasing a tower, with the objective of buying a more performing device and of achieving, even if not needed, lower water temperature, it means the cost for such operation will be higher than planned, and capital will be invested without making a profit. Actual needs may be overrun only if the outcome has been carefully calculated and the need for a considerable water flow tolerance is really advantageous. It is always advisable to contact the evaporative tower builders, as they solve each single problem by employing standard devices with fixed characteristics. It is certainly unlikely that one may precisely offer a tower of the desired capacity, but they will surely recommend a device able to satisfy the desired capacity with a degree of tolerance.

Trustworthy makers do always consider a margin of safety when indicating operating features. One can affirm to really know which is the wet-bulb temperature of a certain locality with good approximation when, for such place, three wet-bulb temperatures for each day, throughout the year, are available, for a period not shorter than 5 or 6 years (this is the minimum; 50 years would be the most acceptable option). For instance, there are studies in the city of Milan which present wet-bulb temperature analysis over the last forty years.

A leading American evaporative tower producer has processed, with an electronic calculator, as many as 10 million observations for 396 localities in the United States and 36 in other countries throughout the world.

Observations were performed every hour, every day, during the hottest period of the year, for ten years. A wet-bulb temperature table was finally drawn up as a basis for designing cooling towers. As the conditions which are most severe for cooling tower operation occur in summer, the table focuses exclusively on the four summer months, when wet-bulb temperatures are at their peak. Operating conditions are more favourable in the other seasons.

The American Society of Heating, Refrigerating and Air Conditioning Engineers (ASHRAE), by examining the data provided by the above mentioned study, prepared a table listing 755 localities in the United States and 310 in other countries.

In addition to the geographical coordinates for each locality, the ASHRAE table features data relating to climatic conditions and, in the last column, the wet-bulb values, as a reference for the planning stage. Such data are divided into three parts with the following headings: 1%, $2\frac{1}{2}$% and 5%. That means that the values they bear may be exceeded in average each year by the 1%, $2\frac{1}{2}$% or 5% of the considered summer hours (2928), which means, respectively, for 30 hours, 75 hours or 150 hours, rounding numbers up.

Of course the temperature in column one is not the maximum temperature, but, as regards tower operation, it is virtually the most severe. As a matter of fact, the 30 hours are only exceptionally consecutive; short periods (2–3 hours) distributed over several days are more likely. Therefore, the thermal flywheel of the plant, which includes the tower, normally enables overcoming such critical points with alterations, which are irrelevant in practice.

An extremely simple practical rule enables reaching the maximum assessable wet-bulb temperature, over a certain period, in a given locality.

For a given locality, the maximum wet-bulb temperature observable in 10 years is the difference between the value in column one (1%) and the one displayed by column two ($2\frac{1}{2}$%). Instead, adding the value in column one (1%) the difference between such value and the one displayed in column three (5%), the result is the maximum value which is probably observable in 50 years.

Of course they are empirical rules, yet they are very close to reality, to such an extent that they are fully acceptable. The question is about deciding which of the three columns contains the best value for each case.

Table 5.2: Wet-bulb temperature in the main European, African, and Asian cities

CITY	1%	2 ½%	5%	CITY	1%	2 ½%	5%
ABADAN	28	27	27	ABIDJAN	28	27.5	27
ADEN	28.5	28	27.5	ADDIS ABEBA	19	18.5	18
ALGIERS	25	24.5	24	AMMAN	21	20.5	20
HAMBURG	20	19	18.5	AMSTERDAM	18.5	18	17
ANKARA	20	19.5	19	ARKHANGELKS	15.5	14.5	14
ASMARA	18.5	18	17.5	ATHENS	22.5	22	22
BAGHDAD	23	22	22	BARCELONA	24	23.5	23
BEIRUT	25.5	25	24.5	BELFAST	18.5	18	17
BELGRADE	23.5	23	22.5	BENGHAZI	25	24.5	24
BERLIN	20	19.5	19	MUMBAI	28	27.5	27
BIRMINGHAM	19	18	17.5	BRUSSELS	21	20	19.5
BUCHAREST	22.5	22	21.5	BUDAPEST	22.5	22	21.5
CAIRO	24.5	24	23.5	KOLKATA	28.5	28	27.5
CARDIFF	18	17.5	17	CASABLANCA	23	22	21
CAPE TOWN	22.5	22	21.5	SINGAPORE	28	27.5	27
COPENHAGEN	20	19	18	SOFIA	22	21	20.5
DAKAR	27.5	27	26.5	JAKARTA	26.5	26	25.5
DUBLIN	18.5	18	17	EDINBURGH	18	17	16.5
JEDDAH	19.5	29	28.5	JERUSALEM	21	20.5	20.5
GIBRALTAR	24.5	24	23.5	GLASGOW	18	17.5	16.5
HANOI	29.5	29.5	29	HANOVER	20	19.5	18.5
HELSINKI	19	18.5	17.5	HONG KONG	27.5	27	26.5
ISTANBUL	24	23.5	23	KATHMANDU	25.5	25	24.5
KHARKOV	20.5	20	19.5	LAGOS	28	27.5	27
SAINT PETERSBURG	18.5	18	17.5	KIEV	20.5	20	19.5
LYON	22	21	20.5	LISBON	20.5	20	19.5
LONDON	20	19	18.5	MADRID	22	20.5	19.5
MANNHEIM	21.5	20.5	20	MARSEILLE	22	21.5	20.5
MILAN	24	23	22	MOGADISHU	28	27	27
MUNICH	20	19	18	MOSCOW	20.5	19.5	18.5
NAIROBI	19	18.5	18.5	NAPLES	23.5	23	22
NICE	23	22	22	NEW DELHI	28.5	28	27.5
ODESSA	21	20.5	20	OSLO	19.5	19	18
PARIS	21	20	19.5	PHNOM PENH	28.5	28	27.5
PRAGUE	19	18.5	18	REYKJAVIK	12	11.5	11.5
ROME	23.5	23	22	THESSALONIKI	25	24.5	24
SAPPORO	24.5	23.5	22.5	SRANNON	18.5	18	17
SHANGHAI	27.5	27	26.5	STOCKHOLM	18	17	16
STRASBOURG	21	20.5	19.5	TEHRAN	24	23.5	23
TEL AVIV	23.5	23	22.5	TOKYO	27.5	27	26.5
TUNIS	25	24.5	23.5	VALENCIA	24	23.5	23
WARSAW	22	21	20	VIENNA	21.5	20.5	19.5
VLADIVOSTOK	21	20.5	20	ZURICH	20	19.5	19

CHAPTER 6

Water Behavior and Treatment in Evaporative Towers

Operativity and performance of evaporating cooling loops may be severely endangered by some specific phenomena, mainly: build-up or incrustation, mud, corrosion.

Such problems can be eliminated or limited by controlling water chemical-physical conditions employing suitable treatments, generally chemical.

6.1 COOLING LOOPS

They are classified into three types:

1. with water from an open system, employing groundwater or superficial water;

2. open, in which cooling water comes in contact with air during each cycle; and

3. closed, in which cooling water does not come in contact with the air but transfers its heat to another fluid inside an exchanger.

   We will now focus exclusively on open loops, namely on point 2.

   Evaporative cooling loops (with evaporative towers or condensers) are open loops, as water comes in contact with air inside the cooling device. They are subject to the issues mentioned above, due to:

- insoluble salts build-up,

- biological growth,

- corrosion of metal parts and alteration of wooden parts,

- mud formation,

- foam formation.

6.1.1 INSOLUBLE SALTS BUILD-UP

Build-up occurs when the concentration of dissolved substances exceeds solubility limit. In the case of evaporative cooling, concentration increase is a natural process as water evaporated through the cooling process is pure, and it is therefore normal that salts tend to build up in the remaining water.

Such condition worsens in presence of substances with inverted solubility characteristics (solubility falls as temperature rises).

The main cause of incrustation is calcium carbonate, which results from calcium bicarbonate decomposition, as per reaction:

$$Ca(HCO_3)_2 = CaCO_3 + CO_2 + H_2O.$$

Other substances which cause build-up are sulphate and calcium phosphate, as well as magnesium and barium salts. Calcium carbonate is soluble in acidified solutions, when incrustation tends to disappear as water pH decreases. However, in many cases water is purposely alkaline (that is with high pH) to reduce corrosion; in such situations calcium carbonate build-up is a common feature.

Langelier saturation index or Ryznar stability index are used, even though for rough calculation, to prevent build-up.

6.1.2   BIOLOGICAL GROWTH

Several types of biological growth may soil different parts of the loop.

Slimy greenish and brown algae grow on surfaces exposed to sun light, dirtying the whole area; if carried by water they may block the system, especially the tower fill pack. Systems which employ marine water may attract mollusks, shells, etc.

Aerobic bacteria owe their name to the fact they take the oxygen they need for surviving directly from the air. Such category includes bacteria and fungi (they belong to the same bacteria class which is useful in biological water depuration plants for clearing water from organic matter). They produce mud that accumulates in every nook of the cooling loop and obstacles, or even makes impossible, both water circulation and air flow inside the tower. Bacteria and fungi can also build up excessive weight on the tower structure. Bacterial flora thick a couple of mm can reach around 1 kg per m^2, and 3–4 mm is its average thickness. Considering that the m^3 of alveolar fill material, which weigh around 30–40 kg, enclose a surface of 200–220 m^2, it is clear that extended biological growth means an overload corresponding to 20 times the fill material. Tower structures are certainly not designed for bearing such loads.

Biological growth is frequent nowadays, especially due to the high degree of water and air organic pollution. In the field of biological depuration, a system which employs a device similar to an evaporative tower with respective fill material, in such case used for depuration purposes, is widely employed.

As fungi and bacteria, safe for rare exceptions, do not need photosynthesis and light for growing, they literally thrive in remote closed parts of the loop.

Some bacteria form a jelly capsule which retains debris, protects them from physical and chemical attacks and promotes mud build-up.

The type of fungi that mainly make up microbiological dirt include string-like and yeast-like fungi, which are associated with other fungi in destructive attacks on the wood parts.

Iron, sulphur and sulphate-reducing bacteria deserve a special mention.

The metabolism of iron bacteria, which are found in fresh water, needs iron; as a consequence iron oxide is deposited on the pipe and equipment walls. They are generally string-like and, if encapsulated, produce extremely bulky mud which dramatically reduces water flow inlet.

Sulphur bacteria, which are able to oxidise sulphur and its compounds, are often present in water containing hydrogen sulphide or generally in sulphurous water.

The string-like types mainly contribute to build-up dirt, and some of the species also trigger a corrosion process by producing sulphuric acid.

Both iron and sulphur bacteria are aerobic and hence grow well in highly aerated water circulating in open loops.

Anaerobic bacteria survive without contact with air by taking oxygen from the compounds which contain it (they therefore deoxidize them, namely they reduce them, hence the definition "reducing"). This kind includes sulphate-reducing bacteria which, by taking oxygen from sulphur compounds that contain it (sulphates and sulphites), produce hydrogen sulphide, which smells bad and corrodes metals.

6.1.3   CORROSION

Electrolytic corrosion risk is a common feature in hydraulic loops in which water comes in contact with different metals.

The corrosion process is accelerated by:

a) low pH value;

b) increase of salts concentration in the water; and

c) increasing cathodic potential difference between metals.

According to cathodic order, the most common metals can be listed as follows:

- zinc;

- aluminium and aluminium alloys;

- mild steel, low-alloy steel, cast iron;

- lead;

- nickel;

- copper alloys; and

- corrosion resistant steel.

Between two metals, the metal which is listed first is the one which undergoes corrosion and hence, the more distant two metals are from each other, the bigger the potential difference between them. Also some microorganisms like sulphur bacteria (aerobic), or sulphate-reducing (anaerobic), participate in metal corrosion. If present, wood parts in towers undergo physical attack caused by the mechanical impact of water, chemical attack (or delignification) due to free chlorine (beyond the ppms), and high concentration of sodium salts and other pH values. Wood chemical treatment does not guarantee protection from microbiological attacks.

6.1.4   MUD

Open hydraulic loops gather pollutants (dust, processing material from make-up water). Such material forms build-up that is generally classified as mud, which is made up of a combination of organic and inorganic matter which tends to soil the fill pack and other parts of the tower loop. Consistency ranges from that of an extremely hard paste to that of thin sand.

6.1.5   FOAM FORMATION

When water falls inside the tower it undergoes strong ventilation and tends to release air, which then forms bubbles. Some of the contaminants tend to stabilize such bubbles generating foam. Foam formation mainly causes circulation pumps to stop due to cavitation, as well as concentrating matter which promotes build-up.

6.2   CHEMICAL CLEANING SYSTEMS

Limescale fouling is mainly dissolved using special organic descaling products.

Other products also contain passivants and can be used for removing limescale in plants including galvanized parts.

There are also products which attack old or petrified build-up, even containing rust or organic pollutants like algae, mud, etc.

Specific corrosion-free safe products are used for opposing algae growth in evaporative water cooling plants.

For cleaning closed-loop piping it is advisable to use special pumps, which are compatible with the descaling products, thus avoiding the waste of time that disassembling the plants to be cleaned would involve.

6.3   WATER PREVENTIVE TREATMENT

A series of specific products can be added to water to stabilize hardness while ensuring plant protection from corrosion.

Some products can be used with superficial temperatures up to 2500°C without prejudice to their quality.

Overdosage is efficient for detaching pre-existing oxide or limescale build-up from the system; underdosage is advisable for limiting limescale build-up to flakes, which can be eliminated by draining the loop, avoiding crystal limescale formation.

Chemical products must not contain humid phosphates in order to not trigger algae growth and not disrupt drainage biological flora.

6.4   PREVENTIVE REMEDIES

Prevention is the best remedy, especially in the case of build-up and corrosion. It involves:

- adequate tower design and

- careful material selection.

Plastic proves extremely suitable in cooling towers for fighting corrosion, as it endures chemical, physical and microbiological attacks quite well, as well as resisting against the reagents employed for opposing such problems.

Another preventive remedy consists of adjusting water treatment to the actual operating conditions of the tower, following careful identification and assessment (water and air flow, temperature, spray loss, drainage, analysis of make-up water, especially involving pH, limescale hardness, total hardness, chlorides, phenolphthalein and methyl orange alkalinity; in some cases also the silica phosphates and iron contents should be determined).

Normalized methods are available for such determinations, as well as simple testing means that anybody can use.

In order to assess the conditions of the loop and prevent problems it is advisable to check circulating water periodically.

6.5   CONCLUSIONS

All the material illustrated in this part is aimed to give installators or users an overview of all the possible issues and advice on how avoiding situations that would hamper the plant operating conditions.

Whenever a problem occurs, the best option is to call upon a specialized company for advice on the best prevention measures and for purchasing the suitable products for carrying them out.

Loop and make-up water should be tested minimum once a year for determining at least the following values:

- pH;

- total hardness (in CaCO3 ppms);

- temporary hardness;

- suspended solids; and

- total alkalinity [1].

CHAPTER 7

Zero-Dimensional Model

7.1 INTRODUCTION

The aim of this work is to determine the essential parameters which characterize an evaporative tower under various operating conditions. Now, the information means employed as support should be highlighted before introducing the mathematical model following used.

The computer market currently proposes new procedures (software) for evaporative tower thermo-fluid dynamic analysis through the finite elements method or the finite volume method. However, such programs, though they can now be used on the personal computers which are normally present in designer studios, solve the problem only partly as they imply a great deal of effort by the designers to get accustomed to them and, in addition, often do not allow simple result interpretation through graphic representation.

Yet, almost the majority of personal computer users usually work with "integrated electronic data sheet" type programs featuring graphic capabilities. They are generally considered tools solely reserved to economic-financial analysis, but in reality are extremely effective for solving heat transfer problems and performing thermo-fluid dynamic analysis.

Electronic sheets come in the form of a table in which the elementary entities, or cells, are determined by means of a line index and a column index, analogously to the elements of a bidimensional matrix. Numerical values, formulas or text strings can be entered in each cell. In detail, formulas are entered in a really natural and not algorithmic way, also referring to the contents of other cells. Electronic sheets generally accept at least three reference types for a cell:

- with absolute positioning (line and column of the cell in object);

- with relative positioning (distance in number of lines and of columns from the cell in which the formula is entered); and

- with a name (assigned to the cell in object).

In plus, most electronic sheets allow concurrent work on various archives. Such feature may be very useful in the case of recursive formulas, for instance those which are obtained thorough the method of the finite differences for the approximated solution of partial derivative differential equations.

7.2   DESCRIPTION OF THE MODEL OF A COUNTERFLOW EVAPORATIVE TOWER

Before describing such specific model the symbols and abbreviation in use must be presented for clear understanding.

$A_{s,o}$: total area of water-air interface surface $[\text{m}^2]$

$C_{0,j}$: integration constants (j=1,2,3) in Equations (7.14) and (7.17)

c_p: specific heat with constant pressure $[\text{Jkg}^{-1}\text{K}^{-1}]$

c_w: water specific heat $[\text{Jkg}^{-1}\text{K}^{-1}]$

h: specific enthalpy $[\text{Jkg}^{-1}]$

q_m: mass flow $[\text{Jkg}^{-1}]$

r: latent heat of vaporization $[\text{J*kg}^{-1}]$

r_0: latent heat of water vaporization at 0°C $[\text{J*kg}^{-1}]$

x: specific humidity $[\text{kg}_w\text{*kg}_{da}^{-1}]$

z: zero-dimensional parameter defined in Equation (7.18)

α: convective heat transfer coefficient $[\text{Wm}^{-2}\text{K}^{-1}]$

ϵ: efficiency

σ: mass transfer coefficient $[\text{kg}_{da}\text{m}^{-2}\text{s}^{-1}]$

subscripts:

a: humid air

da: dry air

i: inlet value

o: outlet value

O: total value

V: vapor

WB: wet-bulb point value

w: water

superscripts:

" : referred to saturated air

Introduced parameters:

$$b = \frac{x_w'' - x_{WB}}{\vartheta_w - \vartheta_{WB}} \; : \; \text{gradient of air saturation straight line (Equation (8.1))} \tag{7.1}$$

$$B = \frac{br_{WB}}{c_{p,a}} \; : \; \text{zero-dimensional gradient of air saturation straight line} \tag{7.2}$$

$$Le = \frac{\sigma c_{p,a}}{\alpha} \; : \; \text{Lewis number} \tag{7.3}$$

$$W = \frac{q_{m,w} c_w}{q_{m,a} c_{p,a}} \; : \; \text{water-air thermal capacity ratio} \tag{7.4}$$

$$X = U X_o \; : \; \text{zero-dimensional coordinate} \tag{7.5}$$

$$X_o = \frac{\alpha A_{s,o}}{q_{m,a} c_{p,a}} \; : \; \text{number of transfer units} \tag{7.6}$$

$$\varepsilon_w = \frac{\vartheta_{w,i} - \vartheta_{w,o}}{\vartheta_{w,i} - \vartheta_{WB}} \; : \; \text{evaporative tower efficiency} \tag{7.7}$$

$$\theta = \frac{\vartheta - \vartheta_{WB}}{\vartheta_{a,i} - \vartheta_{WB}} \; : \; \text{zero-dimensional temperature} \tag{7.8}$$

$$\xi = \frac{x - x_{WB}}{x_{WB} - x_{a,i}} \; : \; \text{zero-dimensional specific humidity.} \tag{7.9}$$

In an evaporative tower water is cooled by the evaporation of part of the same present in the air. Such cooling effect is promoted or reduced by concurrent convective heat transfer between water and air. As no other factors take part in the process, the same is an adiabatic process.

The accuracy of the differential equations that describe the process requires a long numerical procedure for solving the problem. If computers are not used, the only way for obtaining correct results in a reasonable time implies equation simplification. Yet, as a consequence, the degree of accuracy falls.

The first simplified model was introduced by Merkel in 1925. Considering that Lewis number was equal to 1, and disregarding some of the terms of differential equations, he obtained a simple model in which enthalpy difference is the leading term. The possibility to represent such equations graphically with an h, θ diagram made this model rather comprehensible and popular. It was the only model in use for many years, despite its relative inaccuracy and the ambiguity regarding outlet air enthalpy. Some alterations were introduced later in order to obtain a more accurate solution procedure. Merkel's intuition was therefore generally accepted, to such an extent that its applications then became the base for further studies on evaporative towers.

With the appearance of high performance computers, simplifications were no longer necessary as differential equations could be solved numerically; yet the result accuracy issue lives on given the high number of parameters involved. At any rate, Merkel's model is still considered brilliant and widely employed.

This study follows another path for analyzing the adiabatic evaporative process which takes place in the towers. The approach is presented according to the so called zero-dimensional model. The primary objective of such model is the simplicity of the relations between operating parameters, for offering a clear understanding of evaporative towers behavior. In order to achieve the simplest possible model it is assumed that the Lewis number is equal to 1 even though it does not directly appear. This study will focus on counterflow towers in which water falls and air does exactly the opposite (Figure 7.1).

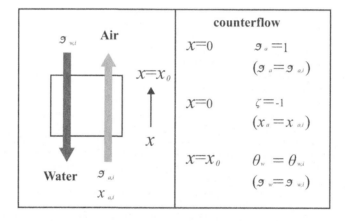

Figure 7.1: Boundary conditions for a counterflow evaporative tower.

The differential equations governing the model are:

$$\frac{d\theta_a}{dX} = -\theta_a + \theta_w \tag{7.10}$$

$$\frac{d\xi_a}{dX} = -\xi_a + B\theta_w \tag{7.11}$$

$$\frac{d\theta_w}{dX} = -\frac{1}{W}\theta_a - \frac{1}{W}\xi_a + \frac{1+B}{W}\theta_w. \tag{7.12}$$

The leading parameters are: temperature difference $(\theta_w - \theta_a)$ for transferring heat (sensible), and the difference of specific humidity $(B\theta_w - \xi_a) = (\xi_w'' - \xi_a)$ for mass transfer between water and air.

The typical equation is 3rd degree with 3 roots:

$$m_1 = 0; \qquad m_2 = \frac{1+B}{W} - 1; \qquad m_3 = -1. \tag{7.13}$$

Root m_2 may be a positive or a negative number. As B and W are independent values, m_2 may be void only by coincidence.

The general solution is (by $m_2 \neq 0$):

$$\theta_a = C_{0,1} + \frac{W}{1+B} C_{0,2} e^{m_2 X} + C_{0,3} e^{-X} \tag{7.14}$$

$$\xi_a = B C_{0,1} + \frac{BW}{1+B} C_{0,2} e^{m_2 X} - C_{0,3} e^{-X} \tag{7.15}$$

$$\theta_w = C_{0,1} + C_{0,2} e^{m_2 X}. \tag{7.16}$$

Given the boundary conditions it is possible to determine constants $C_{0,1}$, $C_{0,2}$ and $C_{0,3}$:

$$\begin{bmatrix} 1 & \frac{W}{1+B} & 1 \\ B & \frac{BW}{1+B} & -1 \\ 1 & e^{m_2 X_0} & 0 \end{bmatrix} * \begin{bmatrix} C_{0,1} \\ C_{0,2} \\ C_{0,3} \end{bmatrix} = \begin{bmatrix} 1 \\ -1 \\ \theta_{w,i} \end{bmatrix}. \tag{7.17}$$

Equations (7.14), (7.15) and (7.16) show the dependence of θ_a, θ_w and ξ_a from X ($0 \leq X \leq X_0$); however, if only their outlet values are needed, or if it is necessary to know the value of efficiency ε_w, a simpler procedure can be developed. The following zero-dimensional parameter is introduced to achieve such objective:

$$z = \frac{1+B}{W} = \frac{q_{m,a} c_{p,a} + q_{m,a} r_{WB} \dfrac{x_w'' - x_{WB}}{\vartheta_w - \vartheta_{WB}}}{q_{m,w} c_w} \tag{7.18}$$

z is the ratio between the total thermal capacity (sensible plus latent) of mass flow given by air following the respective saturation line and the thermal capacity of mass flow given by water. The numerator in Equation (7.18) is fictitious (air does not imply such condition), but it can be easily represented both in an h, x diagram and in the psychrometric one. Through Equation (7.2) parameter B depends on air inlet wet-bulb temperature and on the temperature of the processed water. The W ratio between water mass flow and that of air closely influences z: as a given temperature increases, value z does as well.

A simple formula for calculating efficiency ε_w of an evaporative tower can be obtained by using parameter z:

$$\varepsilon_w = \frac{\vartheta_{w,i} - \vartheta_{w,o}}{\vartheta_{w,i} - \vartheta_{WB}} = 1 - \frac{\theta_{w,o}}{\theta_{w,i}} = z \frac{1 - e^{-(1-z)X_0}}{1 - z e^{-(1-z)X_0}} \tag{7.19}$$

which can be represented in a diagram (Figure 7.2). The effect of the two parameters z and X_0 which influence efficiency is clearly shown. The number of transfer units X_0, as per Equation (7.6), depends on convective thermal transfer coefficient α and on water-air contact area $A_{s,o}$. Equation (7.19) also shows that the limit of ε_w for $X_0 \to \infty$ is $\varepsilon_{w,max} = z$ for $z \leq 1$ and $\varepsilon_{w,max} = 1$ for $z \geq 1$.

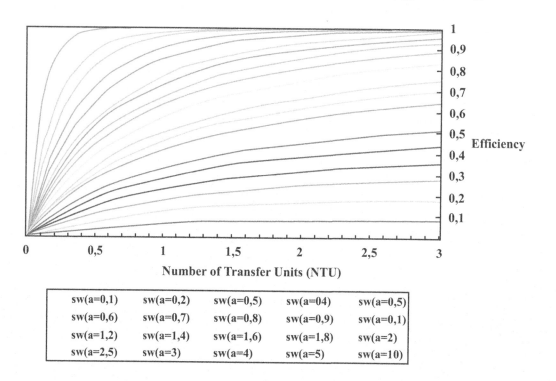

sw(a=0,1)	sw(a=0,2)	sw(a=0,5)	sw(a=04)	sw(a=0,5)
sw(a=0,6)	sw(a=0,7)	sw(a=0,8)	sw(a=0,9)	sw(a=0,1)
sw(a=1,2)	sw(a=1,4)	sw(a=1,6)	sw(a=1,8)	sw(a=2)
sw(a=2,5)	sw(a=3)	sw(a=4)	sw(a=5)	sw(a=10)

Figure 7.2: Water cooling efficiency in an evaporative tower.

The model with $Le = 1$ has an interesting feature: if a new variable $\psi_a = \theta_a + \xi_a$ is entered, the three-equation system (7.10), (7.11) and (7.12) can be split up into two equations:

$$\frac{d\psi_a}{dX} = -\psi_a + (1+B)\theta_w \tag{7.20}$$

$$\frac{d\theta_w}{dX} = \frac{1}{W}[-\psi_a + (1+B)\theta_w] = \frac{1}{W}\frac{d\psi_a}{dX}. \tag{7.21}$$

New variable ψ_a is a zero-dimensional enthalpy of air, more precisely:

$$\psi_a = \theta_a + \xi_a = \frac{h_a - h_{WB}}{c_{p,a}(\vartheta_{a,i} - \vartheta_{WB})}. \tag{7.22}$$

The second term of the right member of Equation (7.20) may be rewritten as follows:

$$(1 + B)\theta_w = \theta_w + \xi_w^{''} = \frac{h_w^{''} - h_{WB}}{c_{p,a}(\vartheta_{a,i} - \vartheta_{WB})}. \tag{7.23}$$

It is clear to see that Equations (7.20) and (7.21) are the zero-dimensional equivalent equations of the celebrated Merkel model, but with a linearized air saturation line. Such equalities also prove that the cooling tower process may, but not necessarily, be expressed as single process operating force solely in function of enthalpy. Hence, with this model no information loss occurs for air temperature and specific humidity as separated values, which instead happens in Merkel's procedure. Equation (7.22) can be used for producing a global energetic balance, valid for all cooling tower types with $Le = 1$:

$$\psi_{a,o} - \psi_{a,i} = \frac{(1 + B)\theta_{w,i}\varepsilon_w}{z} = W(\theta_{w,i} - \theta_{w,o}). \tag{7.24}$$

Such energetic balance implies the assumption of constant $q_{m,w}$ and makes the enthalpy value slightly smaller than what should be air enthalpy inside the tower.

7.3   ADAPTING THE ZERO-DIMENSIONAL MODEL TO THE ACTUAL PROCESS

All the previous equations belong to the zero-dimensional domain and offer a clear comprehension of the effect of the two parameters, z and X_0, which govern the whole procedure. Even though the effect of B (within z) is visible in Figure 7.2, its numerical value is still unknown. As a matter of fact, B is a link between the actual domain and the zero-dimensional one. On the other hand, with (7.1) and (7.2) equations it is clear that B is a function of inlet air wet-bulb temperature θ_{WB}, of water temperature θ_w and of air saturation values. In order to determine efficiency ε_w of an actual evaporative tower, the value of B must be calculated. That means that the relations valid in the zero-dimensional domain must be adapted to the actual air saturation data in order to obtain the final results (dimensional).

The following procedure is adopted for determining the representative value of water temperature θ_w and the values of b and B. It is clear to observe that Equations (7.20) and (7.21), implicitly assuming that air saturation line is straight, can be transformed once again into the dimensional form and be integrated on water-air interface surface $A_{s,o}$ to obtain the following expression:

$$\frac{\sigma A_{s,0}}{q_{m,w}c_w} = \int_0^{A_{s,0}} \frac{dA_s}{q_{m,w}c_w} = \int_{\vartheta_{w,o}}^{\vartheta_{w,i}} \frac{d\vartheta_w}{h_w^{''} - h_a}. \tag{7.25}$$

The same equations (but with the actual air saturation line!) was already proposed in the Merkel model. The base is the idea of an "equivalent" air saturation straight line which could produce the same integral of the right side of Equation (7.25), hence simulating the actual saturation line. The different relation between water temperature and enthalpy h_w'' of saturated air, expressed by the straight and actual air saturation lines, will obviously have an effect on the cooling tower process and consequently cause distribution which is slightly different from air enthalpy h_a along the surface. However, it will be here presumed that in both cases h_a is at any rate function of θ_w. Such two integrals may be substantially simplified:

$$\int_{\vartheta_{w,o}}^{\vartheta_{w,i}} (h_w'')_L d\vartheta_w = \int_{\vartheta_{w,o}}^{\vartheta_{w,i}} (h_w'')_R d\vartheta_w, \tag{7.26}$$

in which L and R, respectively, stand for "linear" and "actual."

Actual enthalpy of saturated air $(h_w'')_R$ is a known function of θ_w (for a given total pressure p) and can be easily integrated. The integral of the right member of Equation (7.26) can be rewritten as follows:

$$\Im = \int_{\vartheta_{w,o}}^{\vartheta_{w,i}} h_w'' d\vartheta_w = \Im(\vartheta_{w,i}) - \Im(\vartheta_{w,o}) \tag{7.27}$$

in which $\Im(\vartheta_{w,i})$ and $\Im(\vartheta_{w,o})$ can be taken from Table 7.1 or calculated using the polynomial coefficients quoted in Table 7.2, for pressure $p = 1$ bar:

$$\Im(\vartheta_w) = a_0 + a_1\vartheta_w + a_2\vartheta_w^2 + a_3\vartheta_w^3 \qquad [kJ \cdot {}^\circ C \cdot kg^{-1}]. \tag{7.28}$$

For pressures which are considerably different from 1 bar, the (7.27) equation can be integrated using appropriate $h_w''(\vartheta_w, p)$ values. In this way the effect or altitude above sea level may be considered according to the pressure in object.

The left side of Equation (7.26) presents $(h_w'')_L$ which represents air enthalpy along the straight air saturation line. Its expression is:

$$(h_w'')_L = c_{p,da}\vartheta_w + [x_{WB} + b(\vartheta_w - \vartheta_{WB})](r_0 + c_{p,V}\vartheta_w). \tag{7.29}$$

Here x_{WB} is the actual specific humidity with wet-bulb inlet air temperature for a given pressure p (or higher).

Replacing the value $(h_w'')_L$ of Equation (7.29) in (7.26) and considering that $\vartheta_{w,m} = (\vartheta_{w,i} + \vartheta_{w,o})/2$ and $c_{p,a} \cong c_{p,da} + x_{WB}c_{p,V}$ (saturated air specific heat) one can obtain an expression for b:

$$b = \frac{\dfrac{\Im(\vartheta_{w,i}) - \Im(\vartheta_{w,o})}{\vartheta_{w,i} - \vartheta_{w,o}} - (x_{WB}r_0 + c_{p,a}\vartheta_{w,m})}{c_{p,V}\left(\dfrac{4\vartheta_{w,m}^2 - \vartheta_{w,i}\vartheta_{w,o}}{3} - \vartheta_{WB}\vartheta_{w,m}\right) + r_0(\vartheta_{w,m} - \vartheta_{WB})}. \tag{7.30}$$

An approximate evaluation of b can be rapidly performed by means of Figure 7.3 by plotting a straight line from point WB' (determined by ϑ_{WB}) on the upper curved line in such way that between

Table 7.1: Numerical values of $\Im(\vartheta_w)$ for total pressure $p = 1$ bar

ϑ_w °C	$\Im(\vartheta_w)$ $kJ \cdot °C \cdot kg^{-1}$	ϑ °C	$\Im(\vartheta_w)$ $kJ \cdot °C \cdot kg^{-1}$	ϑ_w °C	$\Im(\vartheta_w)$ $kJ \cdot °C \cdot kg^{-1}$	ϑ_w °C	$\Im(\vartheta_w)$ $kJ \cdot °C \cdot kg^{-1}$
1	10.425	16	412.635	31	1498.820	46	3886.370
2	22.605	17	459.335	32	1607.670	47	4118.170
3	36.595	18	509.070	33	1722.320	48	4362.820
4	52.450	19	561.980	34	1843.020	49	4620.070
5	70.230	20	618.190	35	1970.070	50	4890.620
6	89.995	21	677.825	36	2103.770	51	5175.270
7	111.800	22	741.040	37	2244.420	52	5474.770
8	135.715	23	807.980	38	2392.370	53	5789.870
9	161.815	24	878.800	39	2548.020	54	6121.470
10	190.180	25	953.675	40	2711.720	55	6470.570
11	220.885	26	1032.790	41	2883.870	56	6838.170
12	254.010	27	1116.330	42	3064.920	57	7225.270
13	289.650	28	1204.495	43	3255.270	58	7632.970
14	327.900	29	1297.490	44	3455.420	59	8062.570
15	368.860	30	1395.520	45	3665.920	60	8515.420

Table 7.2: Polynomial coefficients for Equation (7.28) at $p = 1$ bar

ϑ_w (°C)	a_0	a_1	a_2	a_3
$5 < \vartheta_w \le 20$	-0.672969	10.0723	0.756563	0.0143143
$20 < \vartheta_w \le 40$	-294.945	48.4703	-0.952424	0.040481
$40 < \vartheta_w \le 60$	-7020.16	520.79	-12.0867	0.128689

temperature $\vartheta_{w,o}$ and $\vartheta_{w,i}$, and between the curved line and the straight line, two "triangular" surfaces of identical area appear. The W' and WB' points created with this process are then transferred vertically under the lowest line for determining points W and WB. Hence, a straight line, with gradient obtained from these 2 points, from origin to margin of the diagram determines b.

Such graphic procedure is approximate, as $(h_w'')_L$ in Equation (7.29) is not exactly a linear function of θ_w. Figure 7.4 illustrates the procedure.

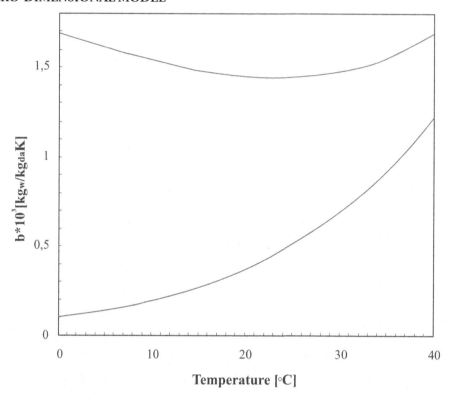

Figure 7.3: Diagram for determining the value of b at $p = 1$ bar.

Of course the intersection between the curve and the air saturation straight line (point W and the one representing water temperature θ_w) is situated between $\vartheta_{w,o}$ and $\vartheta_{w,i}$. Such diagram also reveals the limits of the linearized model:

- linearization of the air saturation curve is introduced assuming that water temperature varies only slightly. If the cooling range $(\vartheta_{w,i} - \vartheta_{w,o})$ is high, deviation of the straight saturation line from the actual one is remarkable and the assumption introduced for Equation (7.26), involving air enthalpy h_a, may be no longer valid. The zero-dimensional model can produce results for each condition, yet it will probably lack accuracy;

- as per Figure 7.4, the distance between points $(WI)_L$ and WB is always shorter than the distance between points WI and WB, so the maximum increase possible for air enthalpy $(h_{a,0} - h_{a,i})_{\max} = (h_{a,WI} - h_{a,i})$ is just a little smaller for the linearized model compared to the actual process. However, a zero-dimensional linear model cannot describe a cooling tower operating in each of the extreme conditions (with minimum air flow), and a process in which inlet air is saturated at water internal temperature $\vartheta_{w,i}$. The application of the linearized

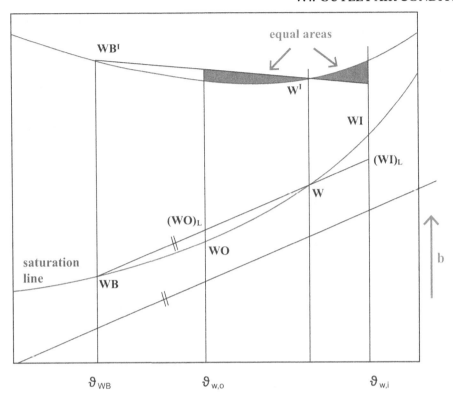

Figure 7.4: Illustration of the graphic procedure for Figure (7.12).

model to such situation would lead to $\varepsilon_W > z$, which is impossible within the zero-dimensional model.

At any rate, none of such two operating conditions are typical in evaporative towers and the merits of the zero-dimensional model should not be judged according to these extreme conditions.

Another limit of the new model must be considered: air is presumed as unsaturated or, in extreme cases, saturated without vaporization. Due to the straight saturation line, air can never generate a nebulised region during the process.

7.4   OUTLET AIR CONDITIONS

Once the values of parameters z, X_o and ε_W are set according to the relations described in the previous paragraph (through iteration if some of temperatures $\vartheta_{w,o}, \vartheta_{w,i}$ or ϑ_{WB} must be determined starting from a given X_o, otherwise through direct procedure if X_o can be calculated knowing temperatures $\vartheta_{w,o}, \vartheta_{w,i}$ and ϑ_{WB}), zero-dimensional external air temperature and specific humidity can be obtained by simply replacing X_o and B values in Equations (7.14) and (7.16), or directly from the following equations:

$$\theta_{a,o} = \frac{\theta_{w,i}\varepsilon_w}{z} + e^{-X_0} = \frac{W}{1+B}\left(\theta_{w,i} - \theta_{w,o}\right) + e^{-X_0} \tag{7.31}$$

$$\xi_{a,o} = \frac{B\theta_{w,i}\varepsilon_w}{z} - e^{-X_0} = \frac{BW}{1+B}\left(\theta_{w,i} - \theta_{w,o}\right) - e^{-X_0}. \tag{7.32}$$

The actual (dimensional) outlet air temperature and specific humidity can be both determined by converting such values with the aid of Equations (7.8) and (7.9), or directly from:

$$\vartheta_{a,o} = \vartheta_{WB} + \left[\frac{1}{z}\left(\vartheta_{w,i} - \vartheta_{w,o}\right) + \left(\vartheta_{a,i} - \vartheta_{WB}\right)e^{-X_0}\right] \tag{7.33}$$

$$x_{a,o} = x_{WB} + \left(x_{WB} - x_{a,i}\right)\left[\frac{B}{z}\frac{\vartheta_{w,i} - \vartheta_{w,o}}{\vartheta_{a,i} - \vartheta_{WB}} - e^{-X_0}\right]. \tag{7.34}$$

Another problem occurs when inlet air is saturated: Equation (7.19) can still be employed, but if $\left(\vartheta_{a,i} - \vartheta_{WB}\right) = 0$ and $\left(x_{WB} - x_{a,i}\right) = 0$, the zero-dimensional values $\theta_{a,o}$ and $\xi_{a,o}$ are undetermined. However, it is possible to calculate the values of $\vartheta_{a,o}$ and of $x_{a,o}$:

$$\vartheta_{a,o} = \vartheta_{WB} + \frac{\varepsilon_w}{z}\left(\vartheta_{w,i} - \vartheta_{WB}\right) \tag{7.35}$$

$$x_{a,o} = x_{WB} + b\left(\vartheta_{a,o} - \vartheta_{WB}\right). \tag{7.36}$$

7.5   ILLUSTRATION OF RESULTS

Two examples may be considered for enhancing comprehension of the zero-dimensional model. The initial data are:

$\vartheta_{w,i} = 32°C$, $\vartheta_{WB} = 20°C$, $\vartheta_{a,i} = 35°C$, $X_0 = 3$, $p = 1$ bar, $\xi_{\alpha,l} = 0.00856\ kg \cdot kg^{-1}$.

The results, displayed by Table 7.3 and represented in Figures 7.5 and 7.6 offer detailed information on the performed process. It must be pointed out that coordinated X in Figure 7.5 is not necessarily proportional to the physical height of the cross section of the tower.

The curve opposing the lines of ϑ_w in Figures 7.5 and 7.6 is a consequence of the opposite sign in the two examples of root m_2 of the typical equation.

The dimensional values calculated within a regular interval of X, are transferred onto a psychrometric diagram (Figure 7.7). The data relating to ϑ_w and x''_w follow air saturation straight line.

7.6   VERIFICATION OF RESULTS

There is no doubt that the best test for verifying the accuracy of the zero-dimensional model involves carrying out a series of field measurements. As such option is not feasible, the results published

Table 7.3: Values calculated for Figures 7.5 and 7.6

Parameter	Example 1	Example 2
$\vartheta_{w,o}$ requested (°C)	27	22
ε_W, Eq. (6-10)	0.417	0.833
z, Eq. (6-10)	0.474	1.205
B, Eq. (A.2)	2.988	2.862
W, Eq. (A.4)	8.421	3.204
x, Eq. (6-29)	1.160	0.843

in various reports can be compared instead. There are two reference sources for performing such process: the exact numerical solutions of the differential equations and the supplier bulletins based upon experimental data which are presumably extremely ample.

Table 7.4 displays a list of results. Columns 11–14 show the values obtained through the zero-dimensional model with Lewis number equal to 1.

N_G value may be easily converted into X_0:

$$N_G = \frac{\sigma A_{s,0}}{q_{m,a}} = \frac{\sigma c_{p,a}}{\alpha} \frac{\alpha A_{s,0}}{q_{m,a} c_{p,a}} = Le X_0 = X_0. \tag{7.37}$$

The results of the zero-dimensional model (column 11) appear quite accurate (better than the Merkel model) for inputs from 0.1–5.3 with extremely low percentage error.

For No. values from 6.1–8.4 results lack accuracy, yet the cooling range ($\vartheta_{w,i} - \vartheta_{w,o}$) is very broad (20 or 30 degrees) and the ($\vartheta_{w,o} - \vartheta_{WB}$) approach is so limited in such examples that linearization obviously introduces an error which is too big.

For No. values equal to 1.1 and 4.1 the new model does not produce results as $\varepsilon_w > z$. Outlet air temperature (respectively 33.62°C and 33.51°C) is too close to inlet temperature (34°C).

However, the examples for No. values from 6.1–8.4 and for Nos. equal to 1.1 or 4.1 correspond exactly to the above mentioned limits of the new model. Therefore, leaving aside such extreme conditions, the linearized model produces acceptable results for X_0.

As regards outlet air conditions $\vartheta_{a,o}$ and $x_{a,o}$, the zero-dimensional results may be generally considered satisfactory. The deviations of the results in columns 13 and 14 from the accurate ones in columns 6 and 7 may be found in the examples marked with an asterisk, but in them the condition of outlet air, calculated according to an accurate model, is nebulised. A formula for unsaturated air enthalpy is used for deriving the zero-dimensional model, errors are hence unavoidable. Nevertheless, outlet air enthalpy is accurately calculated with such model. In addition, external specific humidity is acceptable (even though slightly lower) in these examples, but relative temperature is remarkably lower than what it should be. In general, the external condition obtained in Equations (7.33) and (7.34) should be better checked: if air external temperature and specific humidity highlight nebuli-

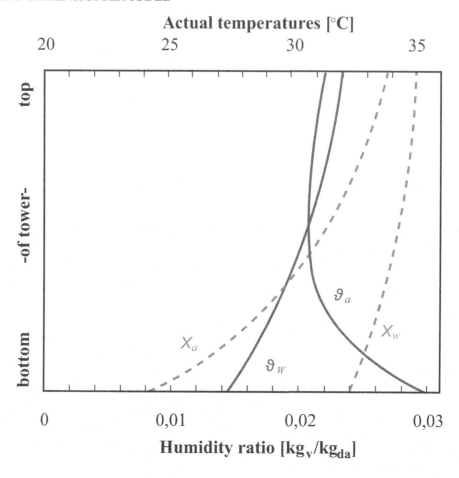

Figure 7.5: Temperatures and specific humidity distribution for Example 1.

sation conditions, outlet air enthalpy and specific humidity can be accepted, while air temperature should be amended.

7.7   OPERATING SIMULATION OF AN EVAPORATIVE TOWER UNDER VARIOUS CIRCUMSTANCES

The difficulty of predicting the performance of a given cooling tower operating under different and variable conditions was solved a priori by plotting the availability and request lines for a number of transfer units on a diagram; the intersection of such two lines generates a new operating point. The procedure is based on the recent computer version of the Merkel model. It is a complicated problem

Table 7.4: Results comparison in an evaporative tower

	Input data					Accurate model			Merkel		Non-dimensional model			
	1	2	3	4	5	6	7	8	9	10	11	12	13	14
No.	$\upsilon_{w,i}$	$\upsilon_{w,o}$	$\upsilon_{a,i}$	$\upsilon_{W,B}$	$\frac{q_{m,a}}{q_{m,w}}$	$\upsilon_{a,o}$	$x_{a,o}$	N_G	N_{GM}	$\Delta\%$	X_O	$\Delta\%$	$\upsilon_{a,o}$	$x_{a,o}$
	°C	°C	°C	°C		°C	g.kg^{-1}				(=N_G)		°C	g.kg^{-1}
0.1	30	26	8	4	0.25	27.01*	23.39*	2.119	1.900	-10.4	2.106	-0.6	26.57	22.73
0.2	30	26	8	4	0.30	24.36*	20.09*	1.396	1.283	-8.1	1.358	-2.7	23.43	19.49
0.3	30	26	8	8	0.30	26.28*	22.61*	1.777	1.615	-9.1	1.746	-1.8	25.30	22.09
1.1	34	30	16	12	0.20	33.62*	34.22*	4.707	3.422	-27.3	-	-	-	-
1.2	34	30	16	12	0.25	30.63*	28.89*	1.861	1.666	-10.5	1.848	-0.7	30.32	28.29
1.3	34	30	16	12	0.30	28.36*	25.29*	1.275	1.667	-8.5	1.247	-2.2	27.90	24.73
1.4	34	30	16	16	0.30	30.49*	28.85*	1.706	1.540	-9.7	1.684	-1.3	29.76	28.37
2.1	34	30	24	20	0.35	32.72	32.45	2.913	2.484	-14.7	3.070	+5.4	32.97	32.10
2.2	34	30	24	20	0.40	31.30	29.75	1.872	1.680	-10.3	1.875	+0.2	31.57	29.40
2.3	34	30	24	20	0.40	30.34	27.71	1.419	1.295	-8.7	1.405	-1.0	30.57	27.40
2.4	34	30	24	24	0.40	32.82	32.66	2.955	2.561	-13.3	3.040	+2.9	32.87	32.42
3.1	34	30	32	28	0.80	32.48	31.05	2.073	1.880	-9.3	2.075	+0.1	32.58	31.02
3.2	34	30	32	28	1.00	32.20	29.45	1.393	1.287	-7.6	1.386	-0.5	32.26	29.42
3.3	34	30	32	28	1.20	32.08	28.36	1.056	0.984	-6.9	1.049	-0.7	32.12	28.35
4.1	34	24	16	12	0.50	33.51	33.97	7.154	5.446	-23.9	-	-	-	-
4.2	34	24	16	12	0.80	27.52*	23.98*	1.564	1.456	-6.9	1.544	-1.3	27.50	23.54
4.3	34	24	16	12	1.00	25.11	20.63	1.086	1.020	-6.1	1.054	-3.0	24.11	20.27
4.4	34	24	16	16	1.00	27.54*	24.23*	1.497	1.397	-6.7	1.444	-3.5	26.90	23.92
5.1	34	24	24	20	1.00	30.07	27.74	2.603	2.404	-7.6	2.534	-2.7	30.38	27.52
5.2	34	24	24	20	1.50	27.66	23.02	1.284	1.211	-5.7	1.223	-4.7	27.88	22.88
5.3	34	24	24	20	2.00	26.65	20.60	0.861	0.817	-5.1	0.817	-5.1	26.80	20.51
6.1	40	20	16	12	1.50	28.21*	25.19*	1.560	1.489	-4.6	1.399	-10.3	27.62	24.90
6.2	40	20	16	12	2.00	25.06*	20.70*	1.031	0.988	-4.2	0.918	-11.0	24.57	20.53
6.3	40	20	16	12	3.00	21.61	16.20	0.617	0.593	-3.9	0.548	-11.2	21.63	16.09
6.4	40	20	16	16	3.00	24.24*	20.02*	0.875	0.839	-4.2	0.745	-14.9	22.75	19.99
7.1	40	20	22	18	3.00	25.85	21.12	1.162	0.127	-3.0	0.978	-15.8	25.96	21.08
7.2	40	20	22	18	5.00	24.18	17.29	0.623	0.606	-2.7	0.530	-15.0	24.23	17.26
7.3	40	20	22	18	8.00	23.32	15.10	0.368	0.358	-2.6	0.314	-14.6	23.34	15.10
8.1	54	24	16	12	1.00	39.55*	49.36*	2.127	2.037	-4.2	1.732	-18.6	37.63	48.90
8.2	54	24	16	12	1.50	33.50*	35.17*	1.150	1.108	-3.7	0.930	-19.1	30.20	35.00
8.3	54	24	16	12	2.00	29.71*	28.13*	0.792	0.764	-3.6	0.640	-19.2	26.57	28.08
8.4	54	24	16	16	2.00	31.70*	37.95*	0.961	0.926	-3.6	0.749	-22.1	27.55	32.11

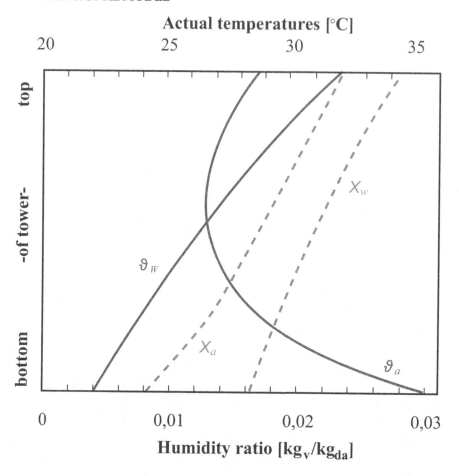

Figure 7.6: Temperature and specific humidity distribution for Example 2.

also for the computer, as X_0 may vary, and the same occurs for other operating parameters such as air and water flow. It is therefore advisable to set X_0 for a given cooling tower with its peculiar geometry and dimension according to water and air flow:

$$X_0 = C \left(\frac{q_{m,a}}{q_{m,w}} \right)^n \tag{7.38}$$

in which C and n are constants determined by at least two known operating points [2].

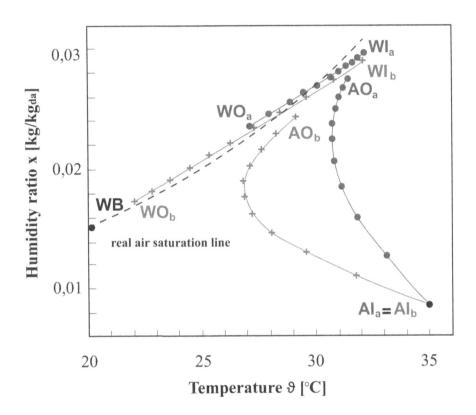

Figure 7.7: Air conditions variation. Examples 1 and 2.

CHAPTER 8

Zero-Dimensional Model Application

To simplify all the operations needed for calculating the variables which play a role in the evaporative tower cooling process, some electronic sheets concerning the zero-dimensional model have been developed. The first sheet named "property" lists the thermophysical properties which are mainly used in the following calculations involving dry air, saturated vapor and liquid water. The relevant proprieties are: density, specific heat, dynamic viscosity, thermal conductivity, Prandtl number, specific humidity and latent heat. The peculiarity of such sheet lies on the fact that all these parameters are exclusively calculated according to temperature under standard atmospheric conditions. In this way it is convenient to act under a mathematical point of view, as the equations are single variable and the variation of the parameters can be easily represented in a diagram. Another strong point is the opportunity to directly compare the properties of dry air, of vapor and of water at the same temperature: hence it is possible to easily obtain all the values that characterize each specific context.

8.1   CALCULATION OF C & n

The second sheet, named "Calculation of C & n," exemplifies a true zero-dimensional model application. Supposing at least two sets of operating conditions are known (Table 8.1), the typical features of an evaporative tower, C and n (see Section 7.7) can be determined.

Wet-bulb temperature, water volume flow rate, cooling range and approach are all data taken from a supplier bulletin. All other values are calculated applying known relations: in detail, the mass flow using density of water at inlet temperature and density of saturated air at wet-bulb temperature.

To calculate the values of b, which is the gradient of the air saturation straight line, in this work it has been preferred to use Equation (7.28) in combination with the 7.2 polynomial coefficient table instead of the graphic method, since one can rely on the electronic sheet. According to the previous paragraph we know that:

$$b = \frac{\dfrac{\Im(\vartheta_{w,i}) - \Im(\vartheta_{w,o})}{\vartheta_{w,i} - \vartheta_{w,o}} - (x_{WB}r_0 + c_{p,a}\vartheta_{w,m})}{c_{p,V}\left(\dfrac{4\vartheta_{w,m}^2 - \vartheta_{w,i}\vartheta_{w,o}}{3} - \vartheta_{WB}\vartheta_{w,m}\right) + r_0(\vartheta_{w,m} - \vartheta_{WB})} \tag{8.1}$$

The first step is determining the two $\Im(\vartheta_w)$. Still using the electronic sheet, the following situation occurs in this case:

Table 8.1: 2 sets of operative conditions of an evaporative tower

Volume flow rate constant ($m^3 \cdot s^{-1}$)		51.82
Operating point	1	2
Wet bulb temperature θ_{WB} (°C)	26	18.95
Air humidity ratio x_{WB} ($kg_w \cdot kg_{da}^{-1}$)	0.02062	0.01344
Specific heat capacity cp_{WB} ($kJ \cdot kg^{-1} \cdot K^{-1}$)	1.8851	1.8787
Cooling range (°C)	12.00	8.00
Approach (°C)	2.55	6.50
Water volume flow rate $q_{v,m}$ ($L \cdot s^{-1}$)	40.06	78.86
Water inlet temperature $\theta_{w,i}$ (°C)	40.55	33.45
Water outlet temperature $\theta_{w,o}$ (°C)	28.55	25.45
Air inlet temperature $\theta_{a,i}$ (°C)	30.00	30.00
Specific heat capacity cp_{da} ($kJ \cdot kg^{-1} \cdot K^{-1}$)	1.0070	1.0070
Water mass flow rate $q_{m,w}$ ($kg \cdot s^{-1}$)	39.74	78.43
Air mass flow rate $q_{m,a}$ ($kg \cdot s^{-1}$)	58.36	60.47

Table 8.2: Polynomial coefficient table

θ_w (°C)	a_0	a_1	a_2	a_3	$(\theta w,i)$ (°C)	$(\theta w,o)$ (°C)	$F(\theta_{w,i})$ (°C)	$F(\theta_{w,o})$ (°C)
							operating point 1	
$5<T\leq20$	-0.673	10.0723	0.756563	0.014314	0	0	-0.672969	-0.672969
$20<T\leq40$	-294.9	48.4703	-0.952424	0.040481	0	**28.55**	-294.945	**1254.5993**
$40<T\leq60$	-7020	520.79	-12.0867	0.128689	**40.55**	0	**2804.2113**	-7020.16
θ_w (°C)	a_0	a_1	a_2	a_3	$(\theta w,i)$ (°C)	$(\theta w,o)$ (°C)	$F(\theta_{w,i})$ (°C)	$F(\theta_{w,o})$ (°C)
							operating point 2	
$5<T\leq20$	-0.673	10.0723	0.756563	0.014314	0	0	-0.672969	-0.672969
$20<T\leq40$	-294.9	48.4703	-0.952424	0.040481	33.45	25.45	1775.811	989.02669
$40<T\leq60$	-7020	520.79	-12.0867	0.128689	0	0	-7020.16	-7020.16

The values in bold are the result of the current situation. Hence, according to initial data it is possible to obtain the other parameter values (Table 8.3) introduced in Chapter 7.

It is now possible to determine the number of transfer units X_0 by using the "search target" function implemented by the electronic sheet, a useful tool for easily performing iterative calculation without having to program algorithms of iterative solutions by oneself. A note for such control is needed to clarify its use: if one knows the desired result of a single formula, but not the input value which generates it, such function varies the value of a specific cell until the formula which depends

Table 8.3: Calculation of zero-dimensional parameters

	r_{WB} [kJ*kg^{-1}]	$c_{p.a}$ [kJ*kg^{-1}*K^{-1}]	b	B	W	z	ε_w
Operating point 1	2440.1	1.046	0.001818	4.2421	2.7192	1.9278	0.8247
Operating point 2	2456.8	1.032	0.001255	2.9872	5.2486	0.7597	0.5517

on it does not return the desired value. One must point out that with the "search target" function one can only modify one cell; it is therefore only suitable for single variable problems. In addition, the explicit value of the target is needed; that means that the target must be assigned a numerical value. The solution is successfully found if the maximum difference between the iterations is under a certain threshold and it stops if the number of interactions exceeds a certain maximum value.

The values obtained for X_0 are then transcribed in Table 8.4.

Table 8.4: Calculation of X_0

	ε_w	$F(X_0)$	X_0
Operating Point 1	0.8247	$1,077.10^{-4}$	1.2747
Operating Point 2	0.5517	$8,588.10^{-4}$	2.0432

Using the equation quoted for convenience $X_0 = C\left(\frac{q_{m,a}}{q_{m,w}}\right)^n$ the required unknown quantities (Table 8.5) are obtained.

Table 8.5: Values of C & n

X_0	$q_{m,a}/q_{m,w}$	n	C
1.2747	1.4685		
		-0.7322	1.6889
2.0432	0.7710		

The values of C and n are really important: they characterize an evaporative tower allowing the classification of several of its aspects under a thermodynamic point of view. In fact, given a cooling tower with its peculiar geometry and dimension the same can be outlined according to the relevant water and air flows.

8.2   CALCULATION OF OUTLET CONDITIONS

The next essential step is to determine water temperature and air conditions when they leave the cooling tower. The sheet "outlet parameter" includes all the mathematical passages needed to achieve such objective. Here are the results obtained for the sample application in object. Supposing to consider the previous tower and assuming also that its operating conditions have been varied due to meteorological changes, the situation is the following (Table 8.6).

Table 8.6: Operating conditions

Volume flow rate constant (m^3*s^{-1})	51.82
Wet bulb temperature θ_{WB} (°C)	21.10
Air humidity ratio x_{WB} ($kg_w*kg_{da}^{-1}$)	0.01535
Specific heat capacity cp_{WB} ($kJ*kg^{-1}*K^{-1}$)	1.8806
Water volume flow rate $q_{v,m}$ ($L*s^{-1}$)	57.41
Water inlet temperature $\theta_{w,i}$ (°C)	42.00
Air inlet temperature $\theta_{a,i}$ (°C)	27.00
Specific heat capacity cp_{da} ($kJ*kg^{-1}*K^{-1}$)	1.0069
Water mass flow rate $q_{m,w}$ ($kg*s^{-1}$)	56.94
Air mass flow rate $q_{m,a}$ ($kg*s^{-1}$)	59.94

As in the previous case, one can determine the parameters needed for developing the zero-dimensional model (Table 8.7).

Table 8.7: Calculation of zero-dimensional parameters

r_{WB} [$kJ*kg^{-1}$]	$c_{p,a}$ [$kJ*kg^{-1}*K^{-1}$]	b	B	W	z	X_o	ε_w
2451.7	1.035	0.001593	3.7708	3.8305	1.2454	1.6266	0.7134

In this case, to evaluate the value of b, without calculating the parameter $\Im(\vartheta_w)$ to make the calculation easier, even though the mathematic procedure in place of the graphic one is once again used for obvious reasons, it is necessary to evaluate the temperature of outlet water a priori. In fact, it is one of our unknown values. It is not difficult to select the initial value: opting for the one which is certainly inferior to the water inlet temperature equal to 10–15°C, it is then possible to calculate the other parameters in this order:

$$B = \frac{b r_{WB}}{c_{p,a}}$$

$$W = \frac{q_{m,w} c_w}{q_{m,a} c_{p,a}}$$

$$z = \frac{1 + B}{W}$$

$$X_0 = C \left(\frac{q_{m,a}}{q_{m,w}} \right)^n$$

$$\varepsilon_w = z \frac{1 - e^{-(1-z)X_0}}{1 - z e^{-(1-z)X_0}}$$

Now, knowing ε_w it is possible to determine the temperature of outlet water:

$$\varepsilon_w = \frac{\vartheta_{w,i} - \vartheta_{w,o}}{\vartheta_{w,i} - \vartheta_{WB}}$$

The only unknown value is precisely $\vartheta_{w,o}$. Comparing the obtained value with the one hypothesized for the calculation of b, if they are different, one can proceed by iteration until these two values converge onto the same temperature rate.

In this case, the third attempt produced a convergence value of:

$$\vartheta_{w,o} = 27,0884°C$$

Instead the supplier bulletin reports $\vartheta_{w,o} = 27,0°C$. That is a 0.327% error, proof of the extraordinary reliability of the employed method.

Air conditions with the data gathered so far can be finally obtained:

$$\vartheta_{a,o} = \vartheta_{WB} + \frac{\varepsilon_w}{z} \left(\vartheta_{w,i} - \vartheta_{WB} \right) = 33,0726°C$$

$$x_{a,o} = x_{WB} + b \left(\vartheta_{a,o} - \vartheta_{WB} \right) = 0,03442746 \ kg \cdot kg^{-1}$$

It is not possible to compare such values with the bulletin, as not released by the supplier, but since they are relations closely correlated to the previous, it is possible to suppose that the error will not be that different (almost identical) from the one obtained for water temperature.

8.3   CALCULATION OF OUTLET AIR ACCORDING TO WATER TEMPERATURE RISE

Another useful value that a planner might need when sizing an evaporative tower is the difference between water inlet and outlet temperature.

For convenience, operating conditions are the same of the tower in the previous paragraph (Table 8.8).

Table 8.8: Cooling tower operating conditions

Volume flow rate constant ($m^3 {*} s^{-1}$)	51.82
Wet bulb temperature θ_{WB} (°C)	21.10
Air humidity ratio x_{WB} ($kg_w {*} kg_{da}^{-1}$)	0.01535
Specific heat capacity cp_{WB} ($kJ {*} kg^{-1} {*} K^{-1}$)	1.8806
Water volume flow rate $q_{v,m}$ ($L {*} s^{-1}$)	57.41
Air inlet temperature $\theta_{a,i}$ (°C)	27.00
Specific heat capacity cp_{da} ($kJ {*} kg^{-1} {*} K^{-1}$)	1.0069
Water mass flow rate $q_{m,w}$ ($kg {*} s^{-1}$)	56.94
Air mass flow rate $q_{m,a}$ ($kg {*} s^{-1}$)	59.94

In this analysis will be supposed the condition: $\theta_{w,i} - \theta_{w,o} = 10$.

The procedure is not different from the one used in the previous paragraph. The only one difference involves the estimation of b: having to hypothesize a temperature for inlet water, the outlet temperature is bound to it as their difference has been chosen as hypothesis. As a consequence, the values of the polynomial coefficient table must be entered with outmost accuracy to avoid mistakes. Firstly $\theta_{w,i}$ is supposed to be equal to 40°C; carrying on with iteration, the temperature values converge with the results shown in Table 8.9.

Table 8.9: Calculation of zero-dimensional parameters

r_{WB} [$kJ {*} kg^{-1}$]	$c_{p,a}$ [$kJ {*} kg^{-1} {*} K^{-1}$]	b	B	W	z	X_o	ε_w
2451.7	1.0358	0.001437	3.4020	3.8306	1.1491	1.6266	0.7134

The temperature values are:

$$\vartheta_{w,i} = 35,827°C$$
$$\vartheta_{w,o} = 25,827°C$$

The supplier bulletin precisely states a temperature of 35.8°C for inlet water. Such approach proves to be really reliable also in this case.

It is now possible to obtain air outlet conditions:

$$\vartheta_{a,o} = \vartheta_{WB} + \frac{\varepsilon_w}{z} \left(\vartheta_{w,i} - \vartheta_{WB}\right) = 29,80°C$$

$$x_{a,o} = x_{WB} + b \left(\vartheta_{a,o} - \vartheta_{WB}\right) = 0,0278609 \text{ kg} \cdot \text{kg}^{-1}$$

8.4   FINAL CONSIDERATIONS

One can observe that the previous values chosen for water and air flow derive from considerations which tend to make the tower operate under optimal conditions: in fact, in principle, their ratio is actually more important than their absolute values. It is clear that in practice the value of $q_{m,w}$ is strongly bound to water distribution mode: values which are too high cause flooding which consequently affects the alveolar fill blocking rising air, while values which are too low enhance the effect of superficial tension. Analogously, $q_{m,a}$ influences the fan power requirements: excessive flow disrupts water laminar flow causing droplet drifting, while a poor flow does not ensure uniform distribution on the fill pack. Evaporating towers have been defined as liquid-gas heat exchangers in which two fluids come into direct contact. The outcome, a close mixture, allows the achievement of heat transfer coefficients higher than those performed by common surface exchangers. Furthermore, it must be highlighted that a role is also played by mass transfer phenomena consequent to differences of the partial pressure of water vapor between the liquid-air interface and the water itself, which enhance the transfer process even further.

The outcome of such heat transfer combined effects is that the minimum temperature that water flowing in the opposite direction of air can in theory reach corresponds to wet-bulb temperature of inlet air, which, as said before, is very close to adiabatic saturation temperature. For this reason nominal project conditions for evaporative towers are assessed in terms of wet-bulb temperature.

Towers can cool water reaching values under ambient air dry-bulb temperature (which is impossible with all processes only based on sensible heat exchange) as wet-bulb temperature is always lower than dry-bulb temperature, except in the case ambient air is saturated.

What occurs in a fully operative system range (Figure 8.1) must be the same that occurs inside the condenser. Range is determined by heat load and water flow, and not by size or capacity of the tower. The approach (closeness to wet-bulb) is the function of the capacity of the tower, hence a larger tower means lower approach (lower outlet water temperature) with identical heat load, flow and air inlet conditions. For all such reasons the typical sizing criteria imply a range of approximately 5–6°C and 10–12°C approach.

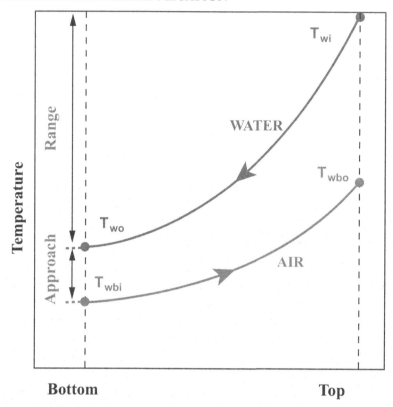

Figure 8.1: Range-approach.

The following observations may be made on the application of the zero-dimensional model in an evaporative tower.

1. Assuming $Le = 1$, water cooling efficiency in a tower can be expressed according just to 2 zero-dimensional independent parameters and their effect in the process becomes rather clear. In addition, efficiency can be calculated in advance and be represented in a ε_w, X_0 diagram with remarkable benefits.

2. The developed procedure is suitable for all kinds of cooling towers. Therefore, it is possibly to carry out a series of comparisons between towers focusing on different aspects, especially efficiency wise.

3. Result accuracy is quite good when considering an application field under moderate conditions. Unfortunately accuracy decreases if the linear model is applied beyond the reasonable limits,

which occurs if the range is too broad; the straight line representation for an actual air saturation line increases the rate of error.

4. The temperature of outlet air and of specific humidity are easy to calculate if considered separately; a feature which is potentially extremely useful for cooling towers, in which outlet humid air density is unknown.

CHAPTER 9

Numerical Analysis

After describing how to determine the external parameters of an evaporative tower with the zero-dimensional model, it is now time to take one step further: the objective is to obtain the partial conditions of water and air in the horizontal sections along the tower. Such problem needs to be solved through numerical analysis. The procedure is based on heat and mass transfer which take place inside the cooling tower. As we will see later, the method is validated by comparison with the results derived from the known experimental data.

Like in the zero-dimensional model chapter, all the involved symbols must be explained.

a_h: area per volume unit of the tower in relation to mass transfer;

a_m: area per volume unit of the tower in relation to heat transfer;

C: specific heat [kJ/kg °C];

F: correction factor for cooled water flow exchange;

g: dry air flow [kg/s];

h: enthalpy [kJ/kg];

h_v: enthalpy of saturated water vapor at temperature t_w [kJ/kg];

h_w: enthalpy of water at temperature t_w [kJ/kg];

L: cooled water flow [kg/s];

Q_T: total transferred heat [kW];

Q_β: transferred latent heat [kW];

r_0: vaporization latent heat of water at temperature t_{ref} [kJ/kg];

RLG: water flow/air flow ratio;

t_w: cooled water temperature [°C];

t_a: temperature of humid air at the same level of temperature t_w [°C];

t_{db}: air wet-bulb temperature [°C];

t_{ref}: temperature of reference for air and water enthalpy, $t_{ref} \approx$ 0.01 °C;

V: tower volume [m^3];

w: specific humidity [kg/kg dry air];

α: heat transfer coefficient for air film [kW/m^2°C];

β: mass transfer coefficient for air film $(w_a - w_i)$ [kg/m^{2}s];

ξ: latent heat/total transferred heat ratio.

Subscripts:

0: state of inlet water and air;

a: humid air;

da: dry air;

i: interface;

n: state of outlet water and air;

v: vapor;

w: water.

9.1 DERIVATION OF THE EQUATIONS

The following assumptions are made:

- Mass and heat transfer occurs only in the air flow direction. Any heat and mass transfer phenomenon between the tower walls and the surrounding environment and heat transfer between the tower fans and the air or water currents is ignored.

- The effects of water loss caused by direction and drainage are insignificant for heat transfer between air and water.

- The cross section areas along the tower are all identical and in each intermediate horizontal section water and air temperature distribution is uniform.

- Mass and heat exchange surfaces are the same, that is $a_h = a_m$.

- The specific heat value of dry air C_{da}, of water vapor C_v and of water C_w are constant. As regards humid air, specific heat is:

$$C_a = C_{da} + wC_v$$

- Heat and mass transfer coefficients α_a, α_w and β are constant inside the tower.

- Lewis number is equal to 1.

A detailed chart of control volume dV in a cooling tower is present in Figure 8.1. The positive direction of dV is from bottom to top along the tower.

Given such hypothesis, under adiabatic flow conditions, the energy conservation equation of the control volume inside the tower is the following:

$$Gdh_a = C_w d\,(t_w L)$$

or

$$Gdh_a = LC_w dt_w + C_w t_w dL. \tag{9.1}$$

This equation affirms that the heat acquired by water equals the heat removed from humid air.

The second term of the right member of Equation (9.1) is usually omitted in a simplified analysis of a standard cooling tower. However, the mass of cooled water dL varies due to the condensation of water vapor in humid air and such quantity must not be disregarded. Therefore, a correction factor is introduced to keep such variation in consideration. By dividing the members of Equation (9.1) by Gdh_a one can obtain:

$$\frac{LC_w dt_w}{Gdh_a} = 1 - \frac{C_w t_w dL}{Gdh_a}. \tag{9.2}$$

By introducing correction factor F in Equation (9.2) one can obtain:

$$dh_a = \frac{LC_w}{GF} dt_w \tag{9.3}$$

in which

$$F = 1 - \frac{C_w t_w dL}{Gdh_a}. \tag{9.4}$$

Equation (9.4) can be manipulated, therefore:

$$F = 1 - \frac{C_w t_w\,(r_0 + h_w)\,dL}{(r_0 + h_w)\,Gdh_a}. \tag{9.5}$$

Latent and total heat are denoted as $dQ_\beta = h_v dL = (r_0 + h_w)dL$ and $dQ_T = Gdh_a$, and considering that $\xi = dQ_\beta/dQ_T$, which is the ratio between latent and total transferred heat, a quantity which may vary according to floor in the tower, Equation (9.5) becomes:

$$F = 1 - \frac{C_w t_w dQ_\beta}{(r_0 + h_w)\,dQ_T} = 1 - \xi\,\frac{C_w t_w}{r_0 + C_w t_w}. \tag{9.6}$$

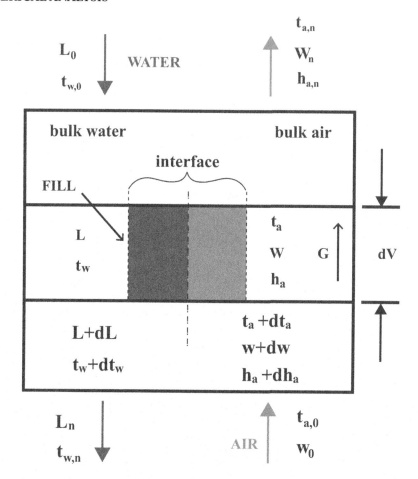

Figure 9.1: Heat and mass transfer in a control volume dV.

Within control volume dV, the quantity of condensed water vapor from the humid air can be expressed as:

$$Gdw = dL = \beta a_m (w - w_i) dV. \tag{9.7}$$

For humid air, water vapor convection and condensation are the heat and mass transfer processes which occur between air-film. We can therefore write:

$$Gdh_a = \alpha_a a_h (t_a - t_i) dV + h_v dL \tag{9.8}$$

in which $h_v = r_0 + h_w = r_0 + C_w t_w$. For the water-film side, water is heated by sensible and latent heat transfer, that is:

$$C_w d(Lt_w) = \alpha_w a_h (t_i - t_w) dV + h_v dL. \tag{9.9}$$

By dividing Equation (9.8) by Equation (9.7) one can obtain:

$$\frac{dh_a}{dw} = \frac{\alpha_a a_h (t_a - t_i)}{\beta a_m (w - w_i)} + h_v. \tag{9.10}$$

Assuming that Lewis number $\alpha_a / \beta C_a = 1$ and that $a_m = a_h$, Equation (9.10) becomes:

$$\frac{dh_a}{dw} = \frac{C_a (t_a - t_i)}{w - w_i} + h_v. \tag{9.11}$$

In accordance with the definition of humid air enthalpy $h_a = C_a t_a + h_v w$, term $C_a (t_a - t_i)$ can be changed into:

$$C_a (t_a - t_i) = (h_a - h_i) - h_v (w - w_i). \tag{9.12}$$

By replacing Equation (9.11) in (9.12) one can find:

$$\frac{dh_a}{dw} = \frac{h_a - h_i}{w - w_i}. \tag{9.13}$$

In addition, with reference to the definition of humid air enthalpy

$$dh_a = C_a dt_a + h_v dw$$

or

$$\frac{C_a t_a}{dw} = \frac{dh_a}{dw} - h_v. \tag{9.14}$$

By combining Equation (9.11) and (9.14) the result is:

$$\frac{dt_a}{dw} = \frac{t_a - t_i}{w - w_i}. \tag{9.15}$$

In which t_i is evaluated as the arithmetic average of corresponding t_a and t_w in dV.
From Equation (9.1)

$$L C_w dt_w = G dh_a - C_w t_w dL \tag{9.16}$$

by dividing both the members of Equation (9.16) by dw and using $dL = Gdw$ the result is:

$$\frac{dt_w}{dw} = \frac{G(dh_a/dw - C_w t_w)}{L C_w}. \tag{9.17}$$

Dry air flow G does not exchange through the tower. On the other hand, the water flow leaving the base of the tower is equal to $L = L_0 + G(w - w_n)$. In such way the Equation (9.17) becomes:

$$\frac{dt_w}{dw} = \frac{(h_a - h_i)/(w - w_i) - C_w t_w}{C_w (L_0/G + w - w_n)}. \tag{9.18}$$

Differential Equations (9.13), (9.15), and (9.18) simultaneously govern the equations for the evaporative tower. They can be solved by numerical integration on the whole volume of the tower, from inlet to outlet air, and by applying to each section the boundary conditions where they can be defined.

9.2 NUMERICAL ANALYSIS GRAPHIC PRESENTATION

Heat and mass exchange between the humid air and the water cooled in a tower can be represented in an h,t diagram. Figure 9.2 illustrates the positions of air state in the cooling and dehumidification process, AB (from tower base to top), the operating line, CD, and "tie line," CE, which are commonly used in classical analysis for cooling water in an evaporative tower. The PQ line represents the enthalpy of saturated air according to air temperature, which means any PQ point represents the enthalpy of the air and the temperature of the interface. Operating line CD, based on Equation (9.3), correlates on the same horizontal section of the tower the air enthalpy area and the area of water temperature. Differently from the straight operating line in the analysis of a standard cooling tower, here CD represents a non-linear relation as it considers the water flow exchange caused by the condensation of the water vapor from humid air. The gradient of section CD, LC_w/GF, varies throughout the tower and reaches the peak at point C, at the base of the tower.

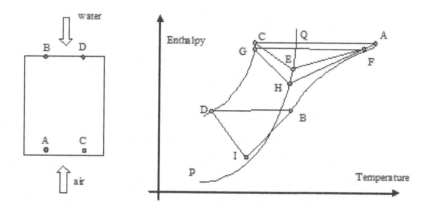

Figure 9.2: h-t diagram in an evaporative tower.

In accordance with Equation (9.1) and for the definition of $\xi = h_v dL/Gdh_a$, Equation (9.9) can be rewritten as follows:

$$Gdh_a = \alpha_w a_h(t_i - t_w)dV + \xi Gdh_a$$

or

$$\alpha_w a_h(t_i - t_w)dV = (1 - \xi)Gdh_a. \tag{9.19}$$

According to numerical analysis, the enthalpy of saturated air could be calculated at the temperature of the interface t_i both of water and of air, rather than at temperature t_w; that is:

$$\frac{dh_a}{h_a - h_i} = \frac{\beta a_m dV}{G}. \tag{9.20}$$

By combining Equations (9.19) and (9.20) one can obtain:

$$\frac{h_a - h_i}{t_w - t_i} = -\frac{\alpha_w}{\beta(1 - \xi)}.$$

(9.21)

Equation (9.21) is generally called "tie line" and represents the quantity of enthalpy transferred through air and the water film. In the diagram the CD tie line connects air enthalpy and water temperature at the interface of air side enthalpy and water side temperature to the same cross section of the tower. Tie line CE gradient, $-\alpha_w/\beta(1 - \xi)$, is also variable, so the tie lines for different sections along the height of the tower are not parallel; for example, CE is not parallel to GH. However, constant tie line gradient is normally applied to the analysis of a standard cooling tower [3].

CHAPTER 10

Numerical Solution Methods

As seen in Chapter 9, a numerical analysis problem can be effectively described with a system of 2nd degree non-linear differential equations; the solutions of a system of such kind only rarely can be written in explicit form, hence the need to find alternative methods for studying their mechanism. A very powerful method is surely simulation through numbers. Numerical simulation of continuous dynamic systems through equations to the differences leads to numerical expressions in which the derivatives are replaced by integration numerical methods. The main issue in numerical simulation is the stability and the accuracy of simulation results when compared with exact ones.

There are various available methods for performing numerical simulation implying different degrees of complexity for carrying them out. As will be shown later, highly complex methods provide better results in terms of stability and of accuracy; the drawback is that simulation time is affected.

10.1 INTRODUCTION

It is necessary to introduce some definitions that will be used later before presenting the methods compared in this chapter.

The objective is finding a numerical solution to Cauchy problem with:

$$X(t) = [x_1(t), x_2(t), x_3(t), \cdots, x_n(t)]^T$$
$$F(t, X(t), U(t)) = [f_1(t, X, U), f_2(t, X, U), f_3(t, X, U), \cdots, f_n(t, X, U)]^T$$
$$U(t) = [u_1(t), u_2(t), u_3(t), \cdots, u_n(t)]^T$$
$$\dot{X}(t) = \left[\frac{dx_1(t)}{dt}, \frac{dx_2(t)}{dt}, \frac{dx_3(t)}{dt}, \cdots, \frac{dx_n(t)}{dt} \right]^T$$

in which $X(t)$ is the vector of the state variables, $U(t)$ is the vector of the inlets.

A generic one-step numerical method is expressed in the following form:

$$X_{n+1} = X_n + T \Psi (T, t_n, X_n) \qquad n = 0, 1, \cdots .$$

In which Ψ depends on function $F(t, X(t), U(t))$ relating to the problem in object and on the method in use, while T is the temporal resolution of the numerical method and remains constant throughout the whole simulation, so that $T = t_{n+1} - t_n$ for each n. Given $X_0 = X(t_0)$ this formula is used for iterative calculation of the values taken on by the state variables at time t_n. Of course such numerical approximation introduces an error at each step, from t_n to t_{n+1}, which can be calculated as the difference between exact value $X(t_{n+1})$ and the theoretical one $\bar{X}_{n+1}$ which is obtained using the numerical method with exact value $X(t_n)$, that is :

$$e_{n+1}^l = X(t_{n+1}) - \bar{X}_{n+1}$$

represents the local truncation error, with

$$\bar{X}_{n+1} = X(t_{n+1}) + T\Psi(T, t_n, X(t_n)).$$

Figure 10.1 is the graphic explanation of the local truncation error.

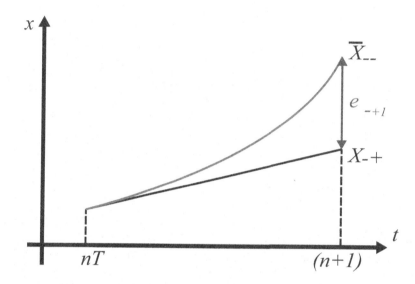

Figure 10.1: Local truncation error.

A method is defined as coherent if the following is true:

$$\lim_{T \to 0} \frac{e_{n+1}^l}{T} = 0.$$

Furthermore, the order of the method can be defined as the biggest positive integer, so the following is true:

$$e_{n+1}^l = O(T^{p+1}).$$

Another important feature of a numerical method is convergence that occurs if the following relation is true:

$$\lim_{T \to 0} X_{n+1} - X(t_{n+1}) = \lim_{T \to 0} e_{n+1}^g = 0$$

in which e_{n+1}^g is the global truncation error at step $n + 1$, namely the error which is made after $n + 1$ steps.

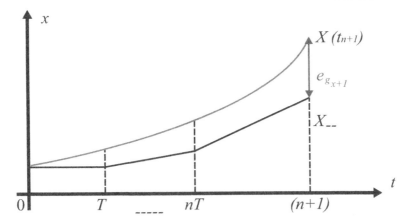

Figure 10.2: Global truncation error.

10.2   EULER METHOD

Euler method is the simplest procedure for approximating a continuous dynamic process to a discrete one. The solutions of the discrete system thus obtained approximate the solutions of the continuous system. The difference between a continuous and a discrete system, namely the discretization error, is a function which rapidly increases as time goes by, hence remarkable calculation power is required for accurate solutions.

To obtain Euler method it is necessary to start by series development $X(t)$ in t_n calculated in $t_{n+1} = t_n + T$:

$$X(t) = X(t) + (t - t_n)X'(t)_{t=t_n} + \frac{(t - t_n)^2}{2}X''(t)_{t=t_n} + \cdots$$

calculated in $t_{n+1} = t_n + T$ returns the following value:

$$X(t_{n+1}) = X(t_n + T) = X(t_n) + TX'(t_n) + \frac{T^2}{2}X''(t_n) + \cdots.$$

It is interrupted at the first derivative, then, bearing in mind that $X'(t_n) = F(t_n, X(t_n))$ and replacing $X(t_n)$ with X_n the result is Euler formula:

$$\begin{cases} X_{n+1} = X_n + TF(nT, Xn) & n = 0, 1, \cdots \\ X_0 = X(t_0) \end{cases}$$

in which for convenience the vector $U(t)$ of the inlets is omitted. It is immediately clear that the local truncation error is an infinitesimal of order higher than the second respect to T, in fact:

$$X(t_{n+1}) = X(t_n) + TF(t_n, X(t_n)) + O\left(T^2\right)$$

therefore Euler is a first-order method.

In brief, Euler method is about updating the value of X_{n+1} by calculating the first derivative in point X_n; Figure 10.3 illustrates such concept.

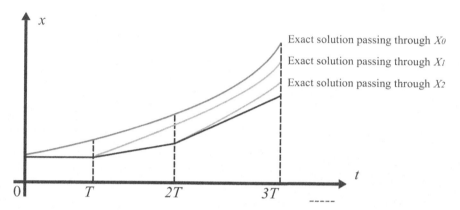

Figure 10.3: Euler numerical method.

10.3 RUNGE-KUTTA METHOD

As seen before, the Euler method is an application of Taylor series development in t_n, calculated on a step of length T; we will now try to calculate it on two $T/2$ length steps:

$$X(t) = X(t) + (t - t_n) X'(t)_{t=t_n} + \frac{(t - t_n)^2}{2} X''(t)_{t=t_n} + \cdots$$

$$X_1 \left(t_n + \frac{T}{2} \right) = X(t_n) + \frac{T}{2} F(t_n, X(t_n))$$

with $X'(t_n) = F(t_n, X(t_n))$, for the first step and

$$X_2 (t_n + T) = X_1 \left(t_n + \frac{T}{2} \right) + \frac{T}{2} F \left(t_n + \frac{T}{2}, X \left(t_n + \frac{T}{2} \right) \right)$$

$$= X(t_n) + \frac{T}{2} F(t_n, X(t_n)) + F \left(t_n + \frac{T}{2}, X \left(t_n + \frac{T}{2} \right) \right)$$

for the full step; also in this case we obtain

$$X(t_n + T) = X_2 (t_n + T) + O \left(T^2 \right)$$

but by duly combining $X_1 \left(t_n + \frac{T}{2} \right)$, $X_2(t_n + T)$ we can eliminate the second-order error, in fact:

$$X(t_n + T) = 2X_2 (t_n + T) - X_1 (t_n + T) + O \left(T^3 \right).$$

By comparing the previous expression with the one in Euler method, it is possible to observe that to gain a unit over the convergence order it is necessary to assess twice the function $F(\cdot)$ for each time interval instead of once. It hence seems possible to obtain higher level algorithms by increasing the number of assessments of function $F(\cdot)$.

Yet it is possible to prove than more than k intermediate time assessments of $F(\cdot)$ are needed for obtaining a convergence order equal to k for $k > 4$ (see Table 10.1); this is why the most advantageous, and most used, calculation algorithm is the order 4 one, which is called classic Runge-Kutta or 4-intermediate-step algorithm.

Table 10.1: Correspondence between order and minimum number of states

Coherence number	1	2	3	4	5	6	7	8	9	10
Minimum number of "s" states	1	2	3	4	6	7	9	11	12 < s < 17	13 < s < 17

The general structure of Runge-Kutta methods is:

$$\begin{cases} X_{n+1} = X_n + T \sum_{i=1}^{s} b_i K_i \\ X_0 = X(t_0) \end{cases}$$

in which s is the number of states and with

$$K_i = F\left(t_n + c_i T, X_n + T \sum_{j=1}^{s} a_{ij} K_j\right) \qquad i = 1, 2, \cdots, s.$$

Parameters a_{ij}, c_i, b_i are real and define the method together with s.

The result of what said above is:

$$\Psi(T, t_n, X_n) = \sum_{i=1}^{k} b_i K_i$$

in addition, considering that a necessary condition for method convergence is

$$K_i = \lim_{T \to 0} K_i = F(t_n, X(t)) \qquad i = 1, 2, \cdots, s.$$

It is now possible to deduce that coherence condition equals to:

$$\sum_{i=1}^{k} b_i = 1.$$

To find the coefficients a_{ij}, c_i, b_i for the fourth-order Runge-Kutta algorithm one must impose the exact solution of differential equations whose solution are fourth-degree polynomials; that guarantees a maximum order method. The coefficients can be effectively represented through Butcher tables. By omitting the demonstration the following values are found:

$$
\begin{array}{c|cccc}
0 & 0 & 0 & 0 & 0 \\
1/2 & 1/2 & 0 & 0 & 0 \\
1/2 & 0 & 1/2 & 0 & 0 \\
1 & 0 & 0 & 1 & 0 \\
\hline
& 1/6 & 1/3 & 1/3 & 1/6
\end{array}
$$

with the layout $\dfrac{c \mid A}{\quad b^T}$.

Hence, the four-stage Runge-Kutta method can be summarized as follows:

$$
\begin{cases}
K_{1,n} = F(nT, X_k) \\
K_{2,n} = F(nT + T/2, X_k + T/2 K_1) \\
K_{3,n} = F(nT + T/2, X_k + T/2 K_2) \\
K_{4,n} = F(nT + T, X_k + T K_3)
\end{cases}
$$

$$
X_{n+1} = X_n + \frac{T}{6}(K_{1,n} + K_{4,n} + 2K_{2,n} + 2K_{3,n}).
$$

With this algorithm function $F(\cdot)$ is evaluated 4 times per time interval, once at start point, twice at the intermediate point and once more at final point, and the four values thus obtained are linearly combined for calculating the new value X_{n+1}, the whole is effectively illustrated in Figure 10.4.

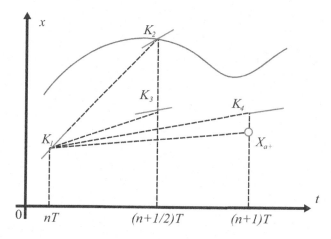

Figure 10.4: Fourth-order Runge-Kutta method.

It is easy to understand that first-order Runge-Kutta method coincides with Euler method if coefficients are chosen in such a way that c=0, a=0 and b=1. As said before, several assessments of

$F(\cdot)$ are required for applying fourth-order Runge-Kutta and hence it appears that a higher degree of accuracy means more complex calculations. Even though such is true in absolute, it is not if algorithm accuracy is the priority. Following an example with an extremely simple scalar case:

$$\frac{dx}{dt} = -x \qquad x(t_0) = 1$$

whose solution is $x(t) = e^{-t}$ and supposing we are interested in the value that variable $x(t)$ takes on at time $t_1 = 10s$, whose exact solution is $x(t_1) = 4.53999 \cdot 10^{-5}$. Table 10.2 summarized the results.

Table 10.2: Comparison between fourth-order R-K and Euler

	R-K	Euler	Euler	Euler
	T = 0.1s	T = 0.1s	T = 0.01s	T = 0.001s
Solution	$4.54003 \cdot 10^{-5}$	$2.65614 \cdot 10^{-5}$	$4.31712 \cdot 10^{-5}$	$4.51733 \cdot 10^{-5}$
$e_g(t_1)/x(t_1)$	$8.81 \cdot 10^{-6}$	0.415	$4.91 \cdot 10^{-2}$	$4.99 \cdot 10^{-3}$
Number of assessments of $F(.)$	400 4 steps per iteration	100 1 step per iteration	1000 1 step per iteration	10000 1 step per iteration

According to the table, although the number of calculations with equal T discretization step is bigger in a Runge-Kutta method, to obtain good results with Euler method, a more accurate degree of discretization is required, which leads to a higher calculation load for the simulator.

10.4   METHODS NUMERICAL STABILITY

A method must necessarily consider convergence for obtaining "good" results, as it focuses on small steps approximation ($T \to 0$); while in practice it is important to use a value for the discretization step which must not be too small in order to avoid time consuming simulation. Hence, the need to understand for which T values discretization errors keep low while calculation proceeds.

Such propriety, which involves the sensitivity of the method to errors, is called numerical stability.

To avoid complications, we can refer to a simple problem in the complex plane:

$$\frac{dx}{dt} = \lambda x$$

with $\lambda = (\alpha + j\omega) \in \mathbf{C}$ and $\alpha < 0$, whose solution, $c \cdot e^{\alpha + j\omega}$ with c a constant that depends on initial conditions. As $\lim_{t \to \infty} x(t) = 0$, it is natural to ask the numerical solution to behave analogously to the continuous solution, which means the following must be true:

$$|X_{n+1}| \le L\,|x_n| \quad \forall n \ge n_0$$

with any n_0, and $0 < L < 1$. Let's concentrate on Euler method:

$$x_{n+1} = x_n + T\lambda x_n = x_n(1 + T\lambda).$$

In order to respect the stability condition, the inequality must be verified

$$|T\lambda + 1| \le 1\ ,$$

hence $T\lambda$ must be inside the unit circle with centre $[-1, 0]$ in the complex plane (Figure 10.5).

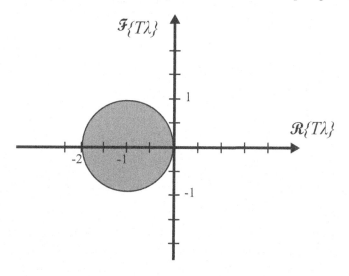

Figure 10.5: Region of absolute stability in Euler method.

The whole can be generalized for any Runge-Kutta method. That means that if we apply a Runge-Kutta algorithm problem $\frac{dx}{dt} = \lambda x$, the result is an equation in the form of:

$$x_{n+1} = x_n Q(T\,\lambda)$$

in which $Q(T\lambda)$ is said stability function. If $|Q(T\lambda)| \le 1$ it means the method is absolutely stable.

By introducing vector $v = (1, 1, 1, 1, 1)^T$ one can demonstrate that the stability function of a Runge-Kutta method is in the form of:

$$Q(T\lambda) = \frac{\det(I - T\lambda A + T\lambda ub^T)}{\det(I - T\lambda A)}$$

which in the case of fourth-order Runge-Kutta specializes in the following formula:

$$Q(T\lambda) = 1 + T\lambda + \frac{1}{2}(T\lambda)^2 + \frac{1}{6}(T\lambda)^3 + \frac{1}{24}(T\lambda)^4.$$

We have so far considered a scalar and linear case, if instead we want to study a dynamic non-linear system with several variables we can equally obtain remarkable results on absolute stability by studying the linearized system around a balance point.

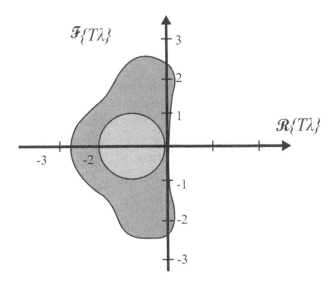

Figure 10.6: Region of asymptotic stability for fourth-order Runge-Kutta method.

Considering the case

$$\frac{dX(t)}{dt} = F(X(t)).$$

Supposing that $\bar{X}$ is a balance point of the dynamic system, meaning that

$$F\left(\bar{X}\right) = \frac{dX(t)}{dt}\bigg|_{X(t)=\bar{X}} = 0$$

then in a surrounding of $\bar{X}$ the following relation is true

$$\dot{X} = F(X) = J_{X=\bar{X}}(X - \bar{X}) + O(X - \bar{X})$$

in which $J_{X-\bar{X}}$ is the Jacobian matrix of $F(X(t))$ calculated at balance point.

$$J = \begin{bmatrix} \frac{\partial f_1}{\partial x_1} & \frac{\partial f_1}{\partial x_2} & \cdots & \frac{\partial f_1}{\partial x_n} \\ \frac{\partial f_2}{\partial x_1} & \frac{\partial f_2}{\partial x_2} & \cdots & \frac{\partial f_2}{\partial x_n} \\ \vdots & \vdots & \ddots & \vdots \\ \frac{\partial f_n}{\partial x_1} & \frac{\partial f_n}{\partial x_2} & \cdots & \frac{\partial f_n}{\partial x_n} \end{bmatrix}.$$

Now, considering eigenvalues of $J\lambda_1, \lambda_2, ..., \lambda_n$; we can say that a numerical method, applied to a non –linear dynamic method is absolutely stable locally if

$$|Q(T\lambda_i)| \leq 1 \qquad i = 1, 2, \cdots, n$$

in which the stability function is the one obtained for a linear one-dimensional system.

It is very interesting to highlight that if we are in the presence of a linear dynamic system, the polynomial peculiar of J, which is, $\det(I - \lambda J)$, is in the same form of the denominator of the transfer function of the system and that therefore the values which the eigenvalues of J take on $\lambda_1, \lambda_2, ..., \lambda_n$, correspond to the poles of the system [4].

CHAPTER 11

One-Dimensional Model Application

11.1 INTRODUCTION

It is beyond doubt that Excel is a fast and easy-to-use spreadsheet application: its appealing interface and multifarious functions, with the integration of Visual Basic for Applications (VBA) programming language, make it a powerful and versatile tool. In addition it is extremely popular and easily accessible.

Some people tend to undervalue the potential of Excel as a calculation tool, even though they already use it for assignments and routine tasks.

To solve the proposed problem through one-dimensional analysis it has hence decided to use VBA functions. In order to program numerical algorithms for solving equations system VBA has been selected because its clearly effectiveness, despite the fact it is time consuming as each algorithm must be both programmed and tested.

In brief these are the advantages which backed such a choice.

- Creation of high quality solutions: a top quality product for easy application implementation.

- Faster solution development: VBA means immediate development in both terms of comprehension and of application speed.

- Reduced implementation costs: the implementation of an application based on a pre-existing one reduces writing and testing for each new code to the minimum, cutting down costs remarkably. Such feature is very useful in the case third persons improve the text in the future.

- The number of samples used in calculation can be easily varied, enhancing result accuracy.

- Fast interaction with the code for changing and carrying out different test sessions on the proposed problem.

11.2 SOLUTION METHOD

In this book the Runge-Kutta method has been adopted. As seen in Section 10.3, it involves considering several terms in Taylor series development function. The problem is that it is necessary to apply further Taylor series developments for evaluating derivatives of higher order. Runge-Kutta method reduces truncation error to a value of $(dw)^5$, whereas value $(dw)^2$ is obtained with Euler method (dw is step size). That means that it is possible to use broader steps while maintaining accuracy. The disadvantage is that there are more calculations for each step; at any rate, by increasing step size, the total calculation load represented by this method should decrease.

For convenience, here are the three differential equations to be solved [(9.13), (9.15) and (9.18)]:

$$\frac{dh_a}{dx} = \frac{h_a - h_i}{x - x_i}$$

$$\frac{dt_a}{dx} = \frac{t_a - t_i}{x - x_i}$$

$$\frac{dt_w}{dx} = \frac{(h_a - h_i)/(x - x_i) - c_w t_w}{c_w(RLG + x - x_n)}.$$

Five initial inputs are needed: four for values at the base of the tower (x_0, $t_{a,0}$, $h_{a,0}$ and $t_{w,n}$) and one at the top of the tower, which is x_n. Such parameters can be obtained using the analytic method mentioned in the previous chapters (zero-dimensional method). After deciding into how many sections the tower must be divided, the values of t_a, h_a and t_w can be determined for each of them.

In addition, it is possible to determine the values of $t_{a,n}$ and $h_{a,n}$ of outlet air and of inlet water temperature $t_{w,0}$ both by means of this method and of the linearized one.

For good result accuracy it is generally advisable to break up the volume of the tower into a number of sections varying from 10–20 units. The number can of course be higher, but in that case calculation speed would be hampered and waiting time extended. The documentation pertinent to VBA source code of the whole program can be found in the Appendix.

11.3 RESULTS ANALYSIS

According to the results obtained for outlet specific humidity w_n and for outlet water temperature $t_{w,n}$ of a tower by employing numerical analysis, air temperature and water temperature can be obtained for each of the cross sections one wish to divide the same into. Tables 11.1, 11.2, 11.3 display calculation examples for different inlet air and outlet water states and for different water-air RLG flow ratios. Since that a large number of sections of the tower is not required for the present purpose, the tower has been divided into 10 sections.

Initial calculation data are entered on the last line of these tables: they are the conditions of inlet air and of outlet water leaving the base of the tower. The first line highlights the conditions of outlet air and inlet water in the upper part of the tower.

Table 11.4 shows a comparison between air and water states obtained through numerical analysis and those obtained through field measurements for different environments and operating conditions. It can be observed that the calculated temperatures and enthalpies of inlet cooled water and the inlet cooled water temperatures are approximately identical to those obtained from the

Table 11.1: Air and water states inside the tower

Section	w_a	t_a	h_a	t_w
	(kg/kg dry air)	(°C)	(kJ/kg)	(°C)
0	0.01129	17.50	46.18	**12.18**
1	0.01114	17.12	45.42	**11.78**
2	0.01099	16.75	44.67	**11.39**
3	0.01084	16.39	43.93	**11.00**
4	0.01069	16.04	43.20	**10.61**
5	0.01054	15.69	42.47	**10.23**
6	0.01040	15.35	41.74	**9.85**
7	0.01025	15.01	41.02	**9.47**
8	0.01010	14.67	40.30	**9.09**
9	0.00995	14.33	39.58	**8.71**
10	0.00980	13.99	38.86	**8.33**

Data: t_{db} =17. 5°C, RLG = 0.45, no. of sections = 10

experimental field data. The percentage of present errors is inferior to 3%, a situation which is considered acceptable. It should be pointed out that only the temperature and enthalpy of outlet air and the temperature of the inlet cooled water are compared as they are parameters which were not adopted as initial inputs.

However, despite the experimental data were obtained by means of an installation on the actual site, it was not possible to compare measurements for air and water in the intermediate sections. Hence, the result of numerical analysis is considered only partly valid. Further experimentation

Table 11.2: Air and water states inside the tower

Section	w_a	t_a	h_a	t_w
	(kg/kg dry air)	(°C)	(kJ/kg)	(°C)
0	0.01479	21.70	59.40	**16.40**
1	0.01445	20.98	57.80	**15.91**
2	0.01411	20.29	56.24	**15.43**
3	0.01378	19.62	54.70	**14.95**
4	0.01344	18.97	53.18	**14.48**
5	0.01310	18.32	51.66	**14.01**
6	0.01276	17.68	50.15	**13.55**
7	0.01242	17.05	48.64	**13.08**
8	0.01208	16.41	47.14	**12.62**
9	0.01174	15.78	45.64	**12.15**
10	0.01140	15.16	44.14	**11.69**

Data: t_{db} =21. 7°C, RLG = 0.76, no. of sections = 10

should be performed in a controlled laboratory in order to obtain full numerical analysis validation by measuring air and water in the intermediate sections inside the tower.

The calculated results could be illustrated graphically for indicating the variation of air and water states inside the tower. For example, Figure 11.1 shows the relation between air specific humidity and air temperature according to Tables 11.1, 11.2, 11.3: a straight line representing the process appears, even if it should be theoretically curved. This is accentuated where a slight air temperature variation is considered (from 14°C–17.5°C, Table 11.1). Furthermore, cooled water flow exchange caused by water vapor condensation is limited, namely the value of F (correction factor, see Equation (9.4)), approximates to the unit. In fact, an optimal calculation value of F

Table 11.3: Air and water states inside the tower

Section	w_a	t_a	h_a	t_w
	(kg/kg dry air)	(°C)	(kJ/kg)	(°C)
0	0.01700	25.50	68.93	**17.63**
1	0.01660	24.51	66.91	**16.81**
2	0.01620	23.67	65.03	**16.05**
3	0.01580	22.87	63.20	**15.31**
4	0.01540	22.10	61.40	**14.58**
5	0.01500	21.34	59.61	**13.85**
6	0.01460	20.59	57.83	**13.13**
7	0.01420	19.85	56.06	**12.41**
8	0.01380	19.11	54.30	**11.69**
9	0.01340	18.38	52.54	**10.97**
10	0.01300	17.65	50.78	**10.25**
Data: t_{db} =25. 5°C, RLG = 0.58, no. of sections = 10				

ranges from 0.9–1.0. Also, three days in summer, which represent a severe condition for the tower, have been taken as reference for analyzing the variation of the fundamental parameters inside the tower according to weather conditions, hour by hour, starting from 8am to 8pm (Figures 11.2, 11.3, 11.4). The whole allowed me to make some observations. Firstly, the representation line is no longer straight but curved due to the reasons illustrated before. Furthermore, prevalently sensible exchange occurs during the hours with lower air temperature, while prevalently latent exchange occurs in higher temperature hours.

Table 11.4: Comparison of calculated and measured parameters

Environment	Parameters	Calculation	Measure	Error (%)
$t_{db} = 17.5°C$, RLG =	$t_{a,n}$ (°C)	13.99	13.8	**1.41**
0.45	$h_{a,n}$ (kJ/kg)	38.86	39.0	**0.36**
	$t_{w,o}$ (°C)	8.33	8.1	**2.88**
$t_{db} = 21.7°C$, RLG =	$t_{a,n}$ (°C)	15.16	15.6	**2.85**
0.76	$h_{a,n}$ (kJ/kg)	44.14	43.5	**1.46**
	$t_{w,o}$ (°C)	11.69	12.0	**2.59**
$t_{db} = 25.5°C$, RLG =	**$t_{a,n}$ (°C)**	**17.65**	**17.8**	**0.83**
0.58	**$h_{a,n}$**	**50.78**	**50.0**	**1.55**
	(kJ/kg)	**10.25**	**10.2**	**0.56**
	$t_{w,o}$ (°C)			

Examining the situation of the average month day for July (black curve) one can see that the behavior is similar to that of a specific day. Therefore, one can affirm that an average month day survey is sufficient for tower performance assessment.

As concerns tower sizing, it is sufficient to refer to the experimental data on fill material behavior to obtain coefficient C and index n of formula

$$X_0 = C \left(\frac{q_{m,a}}{q_{m,w}} \right)^n.$$

Figure 11.5 illustrates air behavior considering humidity variation (from 40%–80%) with inlet air conditions set at 28°C: the blue lines represent a 8°C rise with RLG water-air flow ratio equal to 0.95, the red lines represent a 5°C rise with a RLG ratio equal to 0.5.

In this case, the importance of RLG ratio is clear; as matter of fact with high values of such ratio in the lower part of the tower there is a prevalence of latent exchanges with consequent air cooling, while in the case of low RLG ratio, sensible exchange is prevalent and air temperature still increases from the bottom to the top.

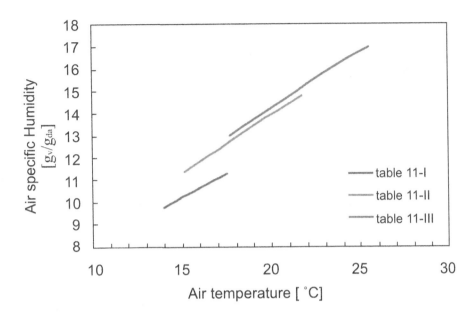

Figure 11.1: Air state variation inside the evaporative tower.

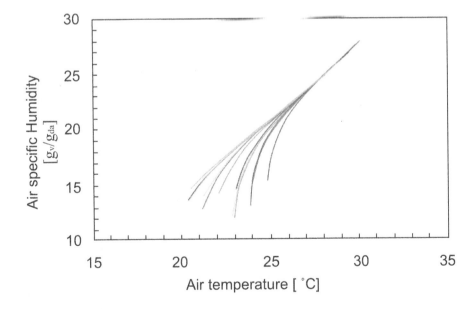

Figure 11.2: Air state variation inside the evaporative tower (June 14th).

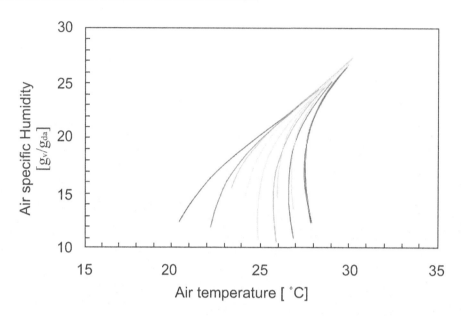

Figure 11.3: Air state variation inside the evaporative tower (July 14th).

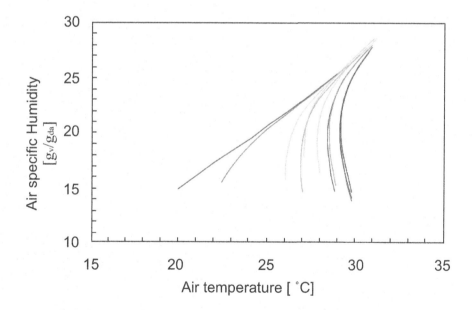

Figure 11.4: Air state variation inside the evaporative tower (August 14th).

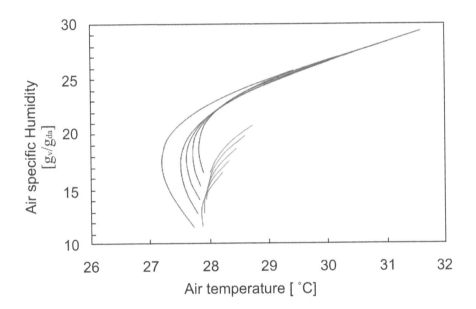

Figure 11.5: Air behavior according to RLG variation.

CHAPTER 12

Conclusions

In this work evaporative towers were studied by employing two simplified methods. Both of them proved sufficiently accurate and the obtained results are in good agreement. In addition, the zero-dimensional method proved its usefulness in determining the operating conditions of fluids to be employed as boundary conditions for the one-dimensional method.

Under moderate conditions the application of the zero-dimensional model is a good combination of simplicity, accuracy and completeness. However, as all linearized models, this particular one is not optimal for describing extreme application conditions, such as remarkable temperature rise between inlet and outlet water, minimum flows and processes in which air nebulises.

As regards the one-dimensional model, further development of the analytic method for evaluating thermal and mass transfer seen in the previous chapters is given by the application of the numerical analysis presented in this work for determining air and water states for each cross section along the tower. The numerical analysis is generally in good agreement with the experimental data, hence the model presented by this work can be deemed extremely efficient. At any rate, if experimental measurements of the intermediate levels of the tower could be carried out and then compared with the results obtained with this method, their degree of reliability, and consequently the reliability of this study, would be further clarified. In addition it could help setting possible alterations for improving the IT code.

This project can be further developed by using more accurate correlations for calculating the parameters which come into play. As a matter of fact, a very large number of variables has not been considered, in order to avoid complicated code layout and consequent longer calculation time.

The creation of a graphic interface for easy data input completes the application.

APPENDIX A

VBA Numerical Code

The following code is used in the VBA environment to determine air and water states for each cross section of an evaporation tower.

```
Public Sub numericalanalysis()
Const k = 273.15

    ' Declares local variables
    Dim ha, hi, ta, ti, w, wi, tw, cw, rlg, wn
    Dim k1(1 To 4), k2(1 To 4), k3(1 To 4), k4(1 To 4), n As
    Integer, dw
    Dim c As Integer


    Application.ScreenUpdating = False

    'enter boundary conditions

    ta = 15.161         'air inlet temperature
    ha = 44.141         'inlet air enthalpy
    tw = 11.693         'inlet water temperature
    w = 0.0114          'inlet air specific humidity
    rlg = 0.76          'water flow/air flow ratio
    wn = 0.01479        'outlet air specific humidity
    n = 10
    dw = (wn - w) / n   'dw humidity difference between outlet and inlet

    'performs the iterations

    For i = 1 To n

        ti = (ta + tw) / 2
        wi = 0.622 * (10 ^ (10.79586 * (1 - k / (ti + k)) + 5.02808 *
            (Log(k / (ti + k)) / Log(10)) + 1.50474 * (10 ^ -4) * _
```

```
        (1 - 10 ^ (-8.29692 * (((ti + k) / k) - 1))) + 4.2873 *
        (10 ^ -4) * (10 ^ (4.76955 * (1 - k / (ti + k))) - 1) _ +
        2.786118312)) / 101325


    hi = (1.045356 * 10 ^ 3 - (0.3161783 * (ti + k)) + 0.0007083814 *
        (ti + k) ^ 2 - 2.705209 * _ (10 ^ -7) * (ti + k) ^ 3) *
        (10 ^ -3) * ti + wi * (2501.6 + (1.3605 * (10 ^ 3) + 2.31334 *
        (ti + k) - 2.46784 _ * (10 ^ -10) * (ti + k) ^ 5 + 5.91332 *
        (10 ^ -13) * (ti + k) ^ 6) * (10 ^ -3) * ti)


    cw = (8.15599 * 10 ^ 3 - 2.80627 * 10 * (tw + k) + 5.11283 *
        10 ^ -2 * (tw + k) ^ 2 - 2.17582 * 10 ^ -13 * (tw + k) ^ 6) *
        10 ^ -3

k1(1) = dw * dhadw(ha, ta, hi, ti, w, wi, tw)
k1(2) = dw * dtadw(ha, ta, hi, ti, w, wi, tw)
k1(3) = dw * dtwdw(ha, ta, hi, ti, w, wi, tw, cw, rlg, wn)

k2(1) = dw * dhadw(ha + k1(1) / 2, ta + k1(2) / 2, hi, ti, w, wi,
        tw + k1(3) / 2)
k2(2) = dw * dtadw(ha + k1(1) / 2, ta + k1(2) / 2, hi, ti, w, wi,
        tw + k1(3) / 2)
k2(3) = dw * dtwdw(ha + k1(1) / 2, ta + k1(2) / 2, hi, ti, w, wi,
        tw + k1(3) / 2, cw,  rlg, wn)

k3(1) = dw * dhadw(ha + k2(1) / 2, ta + k2(2) / 2, hi, ti, w, wi,
        tw + k2(3) / 2)
k3(2) = dw * dtadw(ha + k2(1) / 2, ta + k2(2) / 2, hi, ti, w, wi,
        tw + k2(3) / 2)
k3(3) = dw * dtwdw(ha + k2(1) / 2, ta + k2(2) / 2, hi, ti, w, wi,
        tw + k2(3) / 2, cw, rlg, wn)

k4(1) = dw * dhadw(ha + k3(1), ta + k3(2), hi, ti, w, wi, tw + k3(3))
k4(2) = dw * dtadw(ha + k3(1), ta + k3(2), hi, ti, w, wi, tw + k3(3))
k4(3) = dw * dtwdw(ha + k3(1), ta + k3(2), hi, ti, w, wi, tw + k3(3),
        cw, rlg, wn)


ha = ha + (k1(1) + 2 * k2(1) + 2 * k3(1) + k4(1)) / 6
ta = ta + (k1(2) + 2 * k2(2) + 2 * k3(2) + k4(2)) / 6
```

```
    tw = tw + (k1(3) + 2 * k2(3) + 2 * k3(3) + k4(3)) / 6

    w = w + dw

    ActiveSheet.Cells(i + 1, 1) = w
    ActiveSheet.Cells(i + 1, 2) = ta
    ActiveSheet.Cells(i + 1, 3) = ha
    ActiveSheet.Cells(i + 1, 4) = tw

    Next i
    Application.ScreenUpdating = True

End Sub


Public Function dtadw(ha, ta, hi, ti, w, wi, tw)

    dtadw = (ta - ti) / (w - wi)

End Function


Public Function dhadw(ha, ta, hi, ti, w, wi, tw)

    dhadw = (ha - hi) / (w - wi)

End Function


Public Function dtwdw(ha, ta, hi, ti, w, wi, tw, cw, rlg, wn)

    dtwdw = (((ha - hi) / (w - wi)) - cw * tw) / (cw * (rlg + w - wn))

End Function
```

Bibliography

[1] R. W. Hyland and A. Wexler, Formulations for the Thermodynamic Properties of the saturated Phases of H2O from 173.15K to 473.15K, *ASHRAE Trans.*, 89(2A), 500–519, 1983. 19, 25, 39

[2] Boris Halasz, "Application of a general non-dimensional mathematical model to cooling towers," *Int. Journal of Thermal Science*, 38, pp. 75–88, 1998. DOI: 10.1016/S0035-3159(99)80018-X. 56

[3] Kunxiong Tan and Shiming Deng, "A numerical analysis of heat and mass transfer inside a reversibly used water cooling tower," *Building and Environment*, 38, pp. 91–97, 2003. 75

[4] `etd.adm.unipi.it/theses/available/etd-10142005--121244/unrestricted/capitolo2.doc`: "Analisi numerica," 10/02/2007. 86

[5] Herbert W. Stanford III, "HVAC Water Chillers and Cooling Towers. Fundamentals, Application, and Operation," Stanford White Associates Consulting Engineers, Inc., Raleigh, North Carolina, U.S.A., 2003.

[6] Nicholas P. Cheremisinoff and Paul. N. Cheremisinoff, *Cooling Towers. Selection, Design and Practice*, Ann Arbor Science Publishers, 1981.

[7] Gerald B. Hill, E. J. Pring, Peter D. Osborn, and William Stanford, *Cooling Towers. Principles and Practice*, Butterworth-Heinemann, 1990.

[8] Michele Bianchi, Andrea De Pascale, and Antonio Peretto, *Sistemi Energetici 3. Impatto ambientale*, Ed. Pitagora, 2008.

[9] Gianfranco Coli, *Impianti per la distribuzione dell'energia elettrica negli edifici industriali e civili*, Ed. PEG, Milano, 1993.

Authors' Biographies

ALESSANDRA DE ANGELIS

Alessandra De Angelis received her Ph.D. in Energy Technologies in 2005 from the University of Udine, Italy. She is currently a Research fellow at the Dipartimento di Ingegneria Elettrica, Gestionale e Meccanica, University of Udine. She has taught courses in Thermal Equipments in Buildings within the B.Sc. program in Architecture, Faculty of Engineering and courses in Physics within the B.Sc. program in Agricultural Science and Technology, Faculty of Agriculture. Her research interests include: building thermo-physics, energy saving and renewable energy in buildings, efficiency and energy saving evaluation for heating and cooling plants and computational analysis of thermal and fluid dynamic problems in the commercial equipments; on these topics, she has published many papers in leading journals and conferences.

ONORIO SARO

Onorio Saro is an Associate Professor of Technical Physics at the University of Udine (1999 to present). His scientific activity focuses on the following areas: numerical methods in non linear heat conduction and in heat transfer in conduction-convection coupled problems; computational modeling of free and forced convection; performance of building-plant systems in term of energy requirements and comfort. He has authored over 80 scientific papers, on these topics, published in national and international journals or presented at conferences.

GIULIO LORENZINI

Giulio Lorenzini is Full Professor of Environmental Technical Physics at the University of Parma, Italy. An MSc graduate in Nuclear Engineer (1994) and Ph.D. in Nuclear Engineering (1999) - from Alma Mater Studiorum-University of Bologna, Italy. He is author of some 200 scientific publications, most of which are scientific papers published in international peer reviewed journals and research monographies. He is currently interested in optimization methods in heat transfer and fluid dynamics, Bejan's Constructal Theory, bio-fluid dynamics, analytical and numerical modeling of natural phenomena.

STEFANO D'ELIA

Stefano D'Elia was born in Campobasso in 1985. He received his MSc in Industrial Engineering from the University of Bologna, Italy, with a dissertation on Energy Management. He participated in the VI National Congress of AIGE (2012) illustrating the work on the use of phase change materials (PCM) to recover energy in steel industry. His current interests are in the field of industrial processes optimization in FIAT Group.

MARCO MEDICI

Marco Medici is a Ph.D. student at University of Parma. He holds an MSc in Industrial and Management Engineering. His recent work has focused on Bejan's Constructal Theory applied to heat transfer, water saving modeling in agriculture and energy saving in buildings. He is also the author of some publications in international contexts.

Printed in the United States
by Baker & Taylor Publisher Services